INDUSTRIAL MICROBIOLOGY

INDUSTRIAL MICROBIOLOGY

Shweta Sharma

ANMOL PUBLICATIONS PVT. LTD.
NEW DELHI-110 002 (INDIA)

ANMOL PUBLICATIONS PVT. LTD.
Regd. Office: 4360/4, Ansari Road, Daryaganj,
New Delhi-110002 (India)
Tel.: 23278000, 23261597, 23286875, 23255577
Fax: 91-11-23280289
Email: anmolpub@gmail.com
Visit us at: www.anmolpublications.com

Branch Office: No. 1015, Ist Main Road, BSK IIIrd Stage
IIIrd Phase, IIIrd Block, Bangalore-560 085 (India)
Tel.: 080-41723429 • Fax: 080-26723604
Email: anmolpublicationsbangalore@gmail.com

Industrial Microbiology

First Edition, 2011
ISBN 978-81-261-4516-4

PRINTED IN INDIA

Printed at Mehra Offset Press, Delhi.

Contents

Preface

Industrial microbiology encompasses the use of microorganisms in the manufacture of food or industrial products. The use of microorganisms for the production of food, either human or animal, is often considered a branch of food microbiology. The microorganisms used in industrial processes may be natural isolates; laboratory selected mutants or genetically engineered organisms.

Microbiology began since ancient times when people learned to grind lenses from pieces of glass and combine them to produce magnifications great enough to enable microbes to be seen. Since then microorganisms are exceptionally attractive models for studying fundamental life processes. Microorganisms can be grown conveniently in test tubes or flasks, thus requiring less space and maintenance than larger plants and animals. Microbes grow rapidly and reproduce at an unusually high rate, some species of microbes undergo almost 100 generations in 24 hours period. The metabolic processes of microbes follow patterns that occur among higher plants and animals.

Industrial Microbiology comparatively is a new branch of science which began with the invention of microscope. The present book has been written to meet the specific needs of students of undergraduate and postgraduate classes studying in the colleges and universities of India and abroad. The branch microbiology remained for a long time a discipline almost completely isolated from the rest of biology.

Author

Chapter 1

Microbiological History

Ancient man recognized many of the factors involved in disease. Early civilizations on crete, india, pakistan and scotland invented toilets and sewers; lavatories, dating around 2800 bc, have been found on the orkney islands and in homes in pakistan about the same time. One archaeologist has stated that "the high quality of the sanitary arrangements could well be envied in many parts of the world today". In rome, 315 ad, the public lavatories were places where people routinely socialized and conducted business. Ten to twenty people could be seated around a room, with their wastes being washed away by flowing water; it must have been difficult to *"stand on your dignity"* under such circumstances.

The chinese used toilet paper as early as ad 589. In europe moss, hay and straw were used for the same purpose. I can personally attest to the use as late as 1962 of "slick magazines" as toilet paper in certain european camp grounds. The first cities to use water pipes (of clay) were in the indus valley of pakistan around 2700 bc. Metal water pipes were used in egypt (2450 bc) and the palace of knossos on crete around 2000 bc had clay pipes.

Rome built elaborate aqueducts and public fountains throughout its empire to insure a clean supply of water for its citizens. Rome had a "water commissioner" who was responsible for seeing that the water supply was kept adequate and clean; the punishment for pollution of the water supply was death. Lead was commonly used for roman pipes and the subsequent fall of the roman empire has been related by some to the effects of lead on the roman brain.

Most ancient peoples recognized that some diseases were communicable and isolated individuals thought to carry "infections". An example of this is the universal shunning of lepers, which occurs even today. When the black death struck europe, entire villages were abandoned as people fled in an effort to escape the highly infectious plague.

Similarly, in the middle ages the rich of europe fled to their country homes when small pox struck in an effort to escape its terrible consequences. The fact that people who recovered from a particular disease were immune to that disease was probably recognized many different times in many places. Often these survivors were expected to nurse the ill. Greek and roman physicians routinely prescribed diet and exercise as a treatment for ills.

Sadly, we know that this knowledge did not help most of our ancestors and that the human life span was, until the last 200 years, more often than not cut short due to infectious disease.

Even today approximately 15,000,000 children die per year, mainly from infectious diseases that are preventable with basic sanitation, immunization and simple medical treatments. One might honestly question just how far we have come in our treatment of disease. An excellent synopsis of the history of microbiology ancient people had certainly seen masses of microbes, such as mold and bacterial colonies, on spoiled food, but it is doubtful if anyone considered that they were viewing living organisms.

Small boys and maybe a few love-sick adults staring into a clear pond, must have seen tiny specks moving rapidly about and some may have considered them living creatures, but to express this to their friends would be equivalent to us telling our friends that we'd seen a flying saucer.

The first person to report seeing microbes under the microscope was an englishman, robert hooke. Working with a crude compound microscope he saw the cellular structure of plants around 1665. He also saw fungi which he drew. However, because his lens were of poor quality he was apparently unable to "see" bacteria.

Anton van leeuwenhoek was a man born before his time. Although not the first to discover the microscope or to use magnifying lens, he was the first to see and describe bacteria. We know that he was a "cloth merchant" living in delft holland. And that he used magnifying lens to view the quality of the weave of the merchandise he purchased.

He traveled to england in 1668 to view english cloth and there he saw drawings of magnifications of cloth much greater than any of the current lens available in holland would do.

He returned to holland and took up lens grinding. Being meticulous, he developed his lens grinding to an art and in the process tested them by seeing how much detail he could observe with a given lens.

One can guess that he chanced to look at a sample of pond water or other source rich in microbes and was amazed to see distinct, uniquely shaped organisms going, apparently purposefully, about their lives in a tiny microcosm. He made numerous microscopes from silver and gold and viewed everything he could including the scum on his teeth and his semen.

His best lens could magnify ~300-500 fold which allowed him to see microscopic algae and protozoa and larger bacteria. He clearly had excellent eyesight because he accurately drew pictures of microbes that were at the limit of the magnification of his lens.

He used only single lens and not the compound lens of the true microscopes we employ today; which makes his observations all the more amazing. He wrote of his observations to the royal society of london in 1676 and included numerous drawings.

He astonished everyone by claiming that many of the tiny things he saw with his lens were alive because he saw them swimming purposefully about. This caused no end of shock and wonderment and numerous people hurried to delft to see if this dutchman was "in his cups" or if he was really onto something new and wonderful.

A few minutes with one of his numerous microscopes was all it took to convert his visitors to enthusiastic believers in the existence of these tiny beasties living all around them. His discovery was the equivalent of our finding life on mars today.

SPONTANEOUS GENERATION

The mystery of life has puzzled and confounded humans since the first human began to contemplate his world. The religions of ancient societies were built around the seasons, the sun and the renewal of life as these were so clearly tied to survival; both of the human species through birth and death and of the individual in the attainment of sustenance. Spontaneous generation or the idea that life routinely arises from non-life was a common sense explanation of the miracle of life.

It had the advantage of simplicity, ease of understanding and didn't require any waste-of-time thinking. As science and the scientific method grew with the slow accumulation of knowledge, observant individuals began to consider the origin of life more deeply. Simple observations convinced many people that all the larger animals and plants produced life from previous life. Despite this, the mass of humans clung to the comfortable idea of spontaneous generation further, religions saw it as a convenient way to demonstrate the hand of god operating continuously in the world.

Some individuals, such as j.b. Van helment even described how one could make mice from grain, a jar and dirty rags by putting them together in a dark place for a few weeks and soon mice would appear in the jar.

Other, more perceptive individuals, like f. Redi tested the common idea that maggots arise via spontaneous generation on rotting meat. He placed a piece of meat in three jars, one he left open, one he corked tightly and the third he covered with a fine mesh gauze. Maggots only appeared in the open container, no matter how long he left the jars. Redi's experiment was important because of its eloquent simplicity. Anyone could repeat it and obtain the same clear results. Nevertheless, many people clung fiercely to the idea of spontaneous generation, while others designed experiments to test it.

In every case the results of the majority of these experiments indicated that spontaneous generation did not occur. The intellectual ferment this controversy stirred up gradually evolved into the scientific method as the various antagonists questioned each other's assumptions and, more importantly, their experimental design.

These arguments forced the designing of better experiments and eventually persuaded all but the most recalcitrant believers to discard spontaneous generation as an explanation for all higher life forms.

Then, in the 1800's the refinement of the microscope, through which people could see tiny life forms that they assumed were simple, gave spontaneous generation proponents new life.

Again, flawed common sense led reasonable people astray. As the existence of microscopic life was accepted, the assumption was that such life must be simple compared to higher, more complex life. The reasoning that followed this erroneous assumption was that since the microbes were small they must be simple & it followed that they were formed by spontaneous generation, hence god was still at work creating micro-life. Small is not simple!

The battle over spontaneous generation raged anew both from the pulpit and the lab. A number of scientists performed elementary experiments in which they treated soups and broth's, which left unheated would team with microbes after a few days, with heat to destroy any life present in them and asked the question: "Would new life arise in these sterile soups"? Spallanzani boiled "Soup" in glass containers and melted the glass closed.

The observation that nothing subsequently grew in this "heated" soup suggested that spontaneous generation didn't work.

His detractors, rightly criticized his experiments, proposing that since air is necessary for life and since he had sealed the flask to air, obviously no life could develop. Others boiled soups and microbes grew, thus apparently supporting spontaneous generation. But again the preponderance of data suggested that spontaneous generation did not even apply to the "simple" microbes.

In 1859 one of the fathers of modern microbiology. Pasteur decided to settle the question of spontaneous generation once and for all.

A genius at devising definitive experiments, pasteur first drew the necks of glass flasks out so that they remained open to the air, but were bent so that air could only enter by a curved path. He then added broth and boiled it to destroy contaminating microbes. These flasks were then incubated and observed for months.

He reasoned that the microbes in the air that could contaminate the sterile broth would be trapped on the sides of the thin glass necks before they reached the sterile broth. If spontaneous generation didn't occur no growth should take place. This is exactly what happened, the flasks remained sterile indefinitely, until pasteur tipped the sterile broth up into the curved neck where he predicted the airborne organisms would have settled.

After doing this the broths always grew microbes. These experiments ended the spontaneous generation controversy because these experiment was so elegant and simple and the results so clear, that anyone could repeat them.

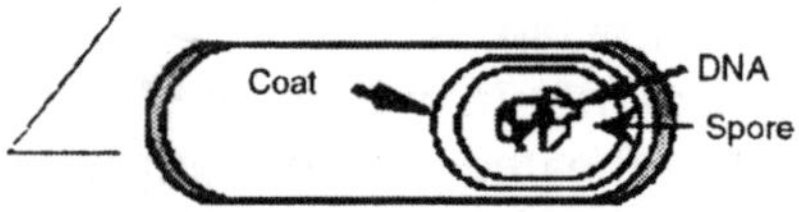

Fig. Spore Structure

Later, an earlier problem, in which occasional heated-broths did not remain sterile, was explained with the discovery of the heat resistant bacterial some of which could can survive several hours of boiling without being killed.

Pasteur discovered many of the basic principles of microbiology and, along with R. Koch, laid the foundation for the science of microbiology. In 1857 Napoleon III was having trouble with his sailors mutinying because their wine was spoiling after only a few weeks at sea. Naturally Napoleon was distraught because his hopes for world conquest were being scuttled (pardon the pun) over a little spoiled wine, so he begged Pasteur for help.

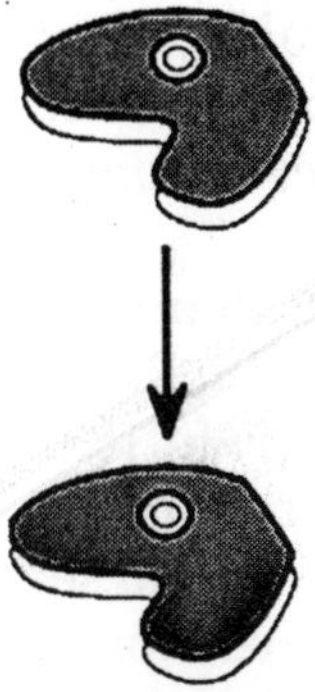

Fig. The Redi Experiment

Pasteur, armed with his trusty microscope, accepted the challenge and soon recognized that by looking at the spoiled wines he could distinguish between the contaminants that caused the spoilage and even predict the taste of the wine solely from his microscopic observations. He then reasoned that if one were to heat the wine to a point where its flavour was unaffected, but the harmful microbes were killed it wouldn't spoil.

As we are aware this process, today known as pasteurization, worked exactly the way he predicted and is the foundation of the modern treatment of bottled liquids to prevent their. It is important to realise that pasteurization is not the same as sterilization. Pasteurization only kills organisms that may spoil the product, but it allows many microbes to survive, whereas sterilization kills all the living organisms in the treated material. Pasteur also realized that the yeast that was present in all the produced the alcohol in wine. When he announced this, a number of famous scientists were enraged, because the current theory of wine production was that wine formation was the result of spontaneous chemical changes that occurred in the grape juice. Pasteur was attacked furiously at scientific meetings, to the point where certain scientists did humorous skits about Pasteur and his tiny little yeast "stills" turning out alcohol. Pasteur had the last laugh however as people all over the world soon realized that if he was right they could control the quality of wine by controlling the yeast that made it. In a short period many others verified his observations and the opposition sank without a sound.

Pasteur discovered many of the basic principles of microbiology and, along with R. Koch, laid the foundation for the science of microbiology. In 1857 Napoleon III was having trouble with his sailors mutinying because their wine was spoiling after only a few weeks at sea. Naturally Napoleon was distraught because his hopes for world conquest were being scuttled (pardon the pun) over a little spoiled wine, so he begged Pasteur for help.

Pasteur, armed with his trusty microscope, accepted the challenge and soon recognized that by looking at the spoiled wines he could distinguish between the contaminants that caused the spoilage and even predict the taste of the wine solely from his microscopic observations. He then reasoned that if one were to heat the wine to a point where its flavour was unaffected, but the harmful microbes were killed it wouldn't spoil.

As we are aware this process, today known as pasteurization, worked exactly the way he predicted and is the foundation of the modern treatment of bottled liquids to prevent their. It is important to realise that pasteurization is not the same as sterilization. Pasteurization only kills organisms that may spoil the product, but it allows many microbes to survive, whereas sterilization kills all the living organisms in the treated material.

Pasteur also realized that the yeast that was present in all the produced the alcohol in wine. When he announced this, a number of famous scientists were enraged, because the current theory of wine production was that wine formation was the result of spontaneous chemical changes that occurred in the grape juice. Pasteur was attacked furiously at scientific meetings, to the point where certain scientists did humorous skits about Pasteur and his tiny little yeast "stills" turning out alcohol. Pasteur had the last laugh however as people all over the world soon realized that if he was right they could control the quality of wine by controlling the yeast that made it. In a short period many others verified his observations and the opposition sank without a sound.

MICROBES

A microbe is any living thing that spends its life at a size visible sometimes only with a microscope. It is too tiny to be seen with the naked eye. Microbes are the oldest form of life on Earth. Some types have existed for billions of years. They may live as individuals or cluster together in communities. Microbes live in the water we drink, the food we eat and the air we breathe.

Right now, billions of microbes are swimming in our belly and mouth and crawling on our skin! Don't worry, over 95% of microbes are good for you. Microbes include bacteria, viruses, fungi, algae and protozoa. These single-cell organisms are invisible to the eye, but they can be seen with microscopes.

Fig. Microbe

The term microbe is short for microorganism, which means small organism. To help people understand the different types of microbes, they are grouped or classified in various ways. Microbes are very diverse and represent all the great kingdoms of life. In fact, in terms of numbers, most of the diversity of life on Earth is represented by microbes.

CLASSIFICATION OF MICROBES

The term microbe is short for microorganism, which

means small organism. To help people understand the different types of microbes, they are grouped or classified in various ways. Microbes are very diverse and represent all the great kingdoms of life. In fact, in terms of numbers, most of the diversity of life on Earth is represented by microbes.

Here is an outline of the major groups of microorganisms:

- Viruses
- Bacteria
- Algae
- Fungi
- Protozoa

Viruses

A virus is too small to be seen without a microscope. A virus is basically a tiny bundle of genetic material carried in a shell called the viral coat. Some viruses have an additional layer around this coat called an envelope.

That's basically all there is to viruses. There are thousands of different viruses that come in many shapes. Many are multi-sided or polyhedral.

If you've ever looked closely at a cut gem, like the diamond in an engagement ring, you've seen an example of a polyhedral shape.

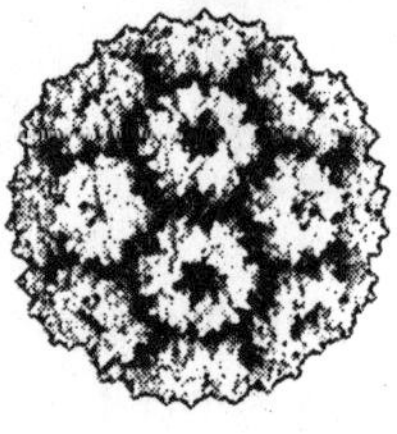

Fig. Viruses

Unlike the diamond in a ring, however, a virus does not taper to a point, but is shaped the same all around. Other viruses are shaped like spiky ovals or bricks with rounded corners. Some are like skinny sticks while others look like pieces of looped string. Some are more complex and shaped like little spaceship landing pods.

Viruses are found on or in just about every material and

environment on Earth from soil to water to air. They're basically found anywhere there are cells to infect. Viruses can infect every living thing.

However, viruses tend to be somewhat picky about what type of cells they infect. Plant viruses are not equipped to infect animal cells, for example, though a certain plant virus could infect a number of related plants.

Sometimes, a virus may infect one animal and do no harm, but cause a great deal of damage when it gets into a different but closely related animal.

Viruses exist to reproduce only. To do that, they have to take over suitable host cells. Upon landing on a suitable host cell, a virus gets its genes inside the cell either by tricking the host cell to pull it inside, or by connecting its viral coat with the host cell wall or membrane and releasing its genes inside, or by injecting their genes into the host cell's DNA.

The viral genes are then copied many times, using the process the host cell would normally use to reproduce its own DNA. The new viral genes then come together and assemble into whole new viruses.

The new viruses are either released from the host cell without destroying the cell or eventually build up to a large enough number that they burst the host cell.

Bacteria

Bacteria consist of only one cell, but they're a very complex group of living things. Some bacteria can live in temperatures above the boiling point and in cold below the freezing point.

There are thousands of species of bacteria, but all of them are basically one of three different shapes. Some are rod- or stick-shaped; others are shaped like little balls. Others still are helical or spiral in shape.

Some bacteria cells exist as individuals while others cluster together to form pairs, chains, squares or other groupings. Some bacteria can make their own food from sunlight, just like plants.

Also like plants, they give off oxygen. Other bacteria

absorb food from the material they live on or in. Some of these bacteria can live off iron or sulfur! The bacteria that live in our stomach absorb nutrients from the digested food you've eaten.

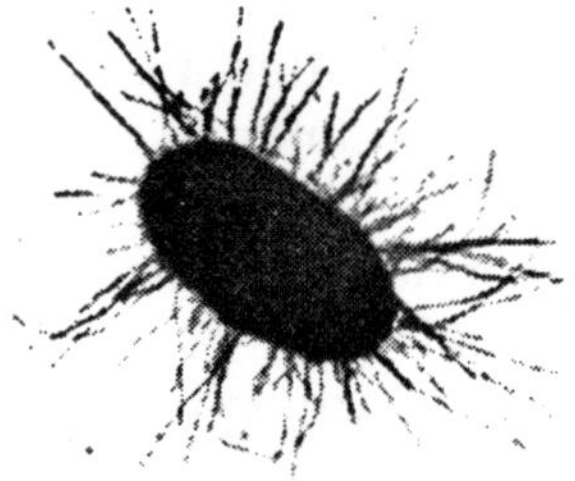

Fig. Bacteria

Some bacteria move about their environment by means of long, whip-like structures called flagella. They rotate their flagella like tiny outboard motors to propel themselves through liquid environments. They may also reverse the direction in which their flagella rotate so that they tumble about in one place.

Other bacteria secrete a slime layer and ooze over surfaces like slugs. Others stay almost in the same spot. Bacteria live on or in just about every material and environment on Earth from soil to water to air and from our body to the Arctic ice to the Sahara deserts.

Each square centimeter of our skin averages about 100,000 bacteria. A single teaspoon of soil contains more than a billion (1,000,000,000) bacteria.

Algae

Algae are found in fresh and salt water around the world. They can also grow on rocks and trees and in soil when enough water is available. Most algae are able to make energy from sunlight, like plants do. They produce a large amount of the oxygen we breathe. However, at some stages of their lives, some algae get their nutrients from other living things.

Diatoms are one kind of algae. They have hard shells made out of glass. When they die, these shells sink to the bottom of their watery environments. We mine deposits of these glass shells that formed hundreds of thousands of years ago to make abrasives, shiny road paint and grit in toothpaste.

Diatoms come in all sorts of shapes—some are round and others are oval. Some look like leaves and others like fat commas.

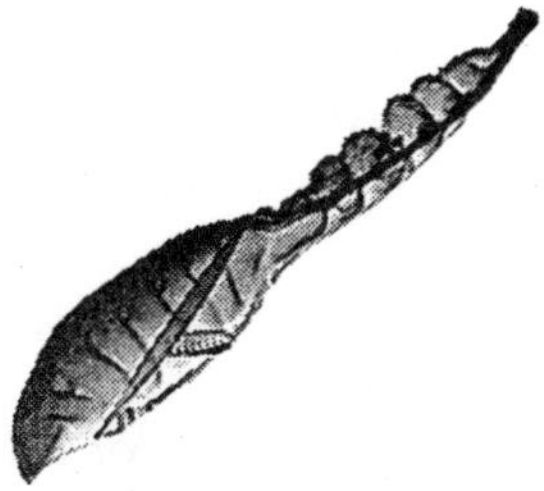

Fig. Algae

Because algae make so much oxygen, these microbes are very helpful. But sometimes certain kinds of algae can also grow in such large numbers that when they suddenly die off en massè, the breaking down of their cells by bacteria destroys the amount of dissolved oxygen in the water, hurting the animals and plants that live there.

Fungi

Fungi are organisms that scientists once confused with plants. However, scientists have found that, at the cell level, the fungi are more like animals than they are like plants.

For one thing, fungi cannot synthesize their own food like plants do, but instead they eat other organisms as do animals. Fungi come in a variety of shapes and sizes and different types. They can range from single cells to enormous chains of cells that can stretch for miles.

Fungi include single-celled living things that exist individually, such as yeast and multicellular clusters, such as molds or mushrooms. Yeast cells look round or oval under a microscope. They're too small to see as individuals, but we can see large clusters of them as a white powdery coating on fruits and leaves.

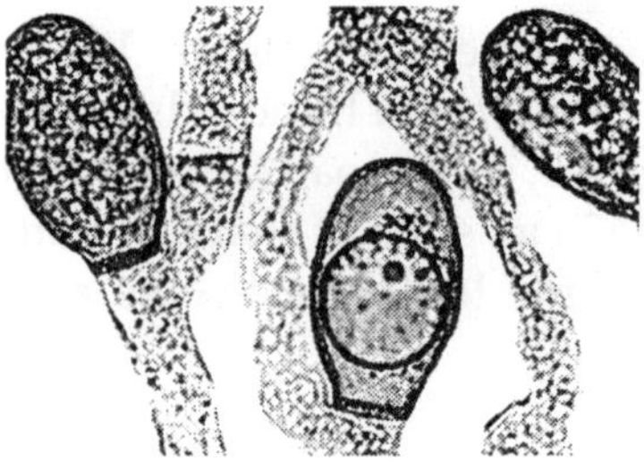

Fig. Fungi

Molds are described as filament-like because they form long filament-like, or thread-like, strands of cells called hyphae. These hyphae are what give mold colonies their fuzzy appearance.

They also form the fleshy body, or mushroom, that some species grow. It may seem strange to think of something as big as a mushroom as a microbe. But the cells of the hyphae making up that mushroom are connected in a closer way than the cells of other multicellular living things, like we and me, are.

Fungi usually grow best in environments that are slightly acidic. They can grow on substances with very low moisture. Fungi live in the soil and on our body, in our house and on plants and animals, in freshwater and seawater. A single teaspoon of topsoil contains about 120,000 fungi.

Fungi are basically stationary. But they can spread either by forming reproductive spores that are carried on wind and rain or by growing and extending their hyphae. Hyphae grow as new cells form at the tips, creating even longer chains of cells. Fungi absorb nutrients from living or dead organic matter that they grow on. They absorb simple, easily dissolved nutrients, such as sugars, through their cell walls. They give off special digestive enzymes to break down complex nutrients into simpler forms that they can absorb.

Some fungi are quite useful to us. We've used several kinds to make antibiotics to fight bacterial infections. These antibiotics are based on natural compounds the fungi produce to compete against bacteria for nutrients and space. We use baker's yeast, to make bread rise and to brew beer. Fungi break down dead plants and animals and keep the world tidier. We're exploring ways to use natural fungal enemies of insect pests to get rid of these bugs.

There are some dangerous fungi that cause diseases in plants, animals and people. Fungi ruin about a quarter to half of harvested fruits and vegetables each year.

MICROBES ARE HELPFUL

Microbes are much more our friends than our enemies. Although some microbes cause health problems such as strep throat, chickenpox and the common cold, most microbes make our lives better such as: Bacillus thuringiensis - a common soil bacterium that is a natural pest-killer in gardens and on crops. Arbuscular mycorrhizas - fungus living in the soil that helps crops take up nutrients from the soil. Accharomyces cerevisiae - baker's yeast that makes bread rise.

Escherichia coli - one of many kinds of microbes that live in our digestive system to help we digest our food every day. Streptomyces - bacteria in soil that makes an antibiotic used to treat infections. Pseudomonas putida - one of many microbes that clean wastes from sewage water at water treatment plants. Lactobacillus acidophilus - one of the bacteria that turn milk into yogurt.

There are many other important jobs microbes do. They are used to make medicine. They break down the oil from oil spills. They make about half of the oxygen we breathe. They are the foundation of the food chain that feeds all living things on earth. We've been using microbes for thousands of years to make products we need and enjoy. For example, we can thank fungi for the cheese on our cheeseburger and yeast for our bun. Cheese and bread are two microbe-made foods people have been enjoying since time began.

In pollution control, researchers are using bacteria that eat methane gas to clean up hazardous waste dumps and landfills. These methane-eating bacteria make an enzyme that can break down more than 250 pollutants into harmless cells.

By piping methane into the soil, researchers can increase growth of the bacteria that normally live in the polluted soil.

More bacteria means faster pollution break up. Also, bacteria is being used as one of the tools to clean up oil spills. These bacteria eat the oil, turning it into carbon dioxide and other harmless by-products. Fungi and bacteria produce antibiotics such as penicillin and tetracycline.

These are medicines we use to fight off harmful bacteria that cause sore throats, ear infections, diarrhea and other discomforts. Scientists have changed the genetic material of bacteria and yeasts to turn them into medicine. They inject genes for medicines they want to make into the microbe cells, as if adding new building information to the microbe's cell DNA. The scientists then grow the microbes in huge containers called fermenters where they reproduce into billions, all making new medicines.

Microbes and the Days of Creation

The world of germs and microbes has received much attention in recent years—and for good reason. We frequently hear the term microbe associated with organisms such as Escherichia coli, Salmonella, anthrax bacteria, antibiotic-resistant tuberculosis, MRSA, HIV, malaria, Stachybotrys and other microscopic creatures. But where do microbes fit into the creation account? Were they created along with the rest of the plants and animals in the first week of creation, or were they created later, after the Fall?

Are microbes a result of the Curse? These and other questions are some that a group of professional creation microbiologists have been asking and their answers may surprise us. Ongoing research based on the creation paradigm appears to provide some answers to these puzzling questions.

Although we cannot be dogmatic about the details of microbe origin during Creation Week, it is believed that a reasonable extrapolation from biological data and Scripture can be made about the nature of microbes in a fully mature creation. Past creation scientists such as Leeuwenhoek, Pasteur and Lister, were blessed by God as He revealed to them critical insight into His creation. So, where do these microbes fit into the very good days of creation? Before answering this question, three terms, microbe, germ and symbiosis, need to be defined. These are relatively new or "modern" terms. First, the Bible does not use these specific terms. These terms were not commonly used until the end of the nineteenth century.

The term microbe was first used in 1878 to describe "extremely minute living beings." Before 1878, scientists including Louis Pasteur, used a variety of terms rather loosely to label the very small organisms that had interested them. It was not clear whether microbes belonged to the animal or plant kingdom, or to a completely different one.

The term microbe was given by Charles E. Sedillot to describe bacteria Later, it would also be used of eukaryotic cells, including algae, fungi, protozoans and slime molds. Some people refer to viruses as microbes, but others do not because viruses are not cells. Viruses have nonliving, as well as living characteristics.

Today, the term germ refers to disease-causing microbes, or pathogens. All germs would have originated after the Fall The Edenic Curse would have profoundly influenced all creation, including viruses, bacteria, fungi and protozoans that would later become pathogens or parasites. The origin of infectious disease is complex and multifaceted. This topic is further explored in the book.

The Genesis of Germs It provides some understanding into the origin of infectious diseases. From a biblical worldview, infectious diseases and patho-genesis are a secondary state in nature. It is not the way the Creator intended for man and nature.

Most microbes are beneficial to man and nature. Only about 5–10% of all bacteria are pathogenic. Many microbes live in a mutualistic relationship with plants, animals, humans and other microbes. Mutualism is a type of symbiosis. The term symbiosis is used to describe an intimate association between organisms of different species. Some symbiotic relationships between microbes and plants, animals and humans are essential for life on earth.

For this reason most microbiologists maintain that bacteria, fungi, protists and other microbes have been maligned in the news media. Without our intestinal flora, we would not digest food nor acquire vitamins and minerals very efficiently. Without fungi, bacteria, algae and protozoans, life on earth could not last. This is because microbes provide essential "services" (e.g., nitrogen fixation, nitrification and denitrification) in nutrient cycles.

Viruses Fit into Creation

The determination of virus origin is uncertain. It may be that viruses have multiple origins. Some may be degenerate parts from cells after the Curse; still others may have their origin during the days of creation. Today, we think of viruses (Latin for "poison") only in the context of disease. However, some viruses (or at least virus-like genes) are involved in a positive function in nature.

Some groups of viruses, like bacteriophages, play a positive role in controlling bacteria in ecosystems and may play a role in diversity. Another group of viruses play a role in turning off the immune system during pregnancy in mammals and humans.

This is a group referred to as endogenous retroviruses (ERVs). ERVs are among a kind of repetitious genetic elements called "retrotransposons". Research has shown that the ERV design prohibits the mother's immune system from damaging the child's body.

These retroviruses cannot fully replicate, only expressed in local immune cells (such as macrophages) of the placenta, thereby preventing them from initiating a full-blown immune response. Thus, the mother's immune system remains competent to respond to other infections but is specifically prevented from mounting an immune response to the developing embryo. So in creation, the selective ability to turn off the immune system for protection would be a "good" design. Other ERVs also play a positive role in animal and human reproduction. However, since the corruption of creation, the corrupted retrovirus, HIV and various leukemia viruses turn off the entire immune system, leaving the body open to devastating infections.

These examples may provide clues to the origin of viruses and how some may have been created during Creation Week by design and how some have been corrupted as a result of the Fall.

Aseptic Technique

Trying to study this mixed population is often difficult and in the tradition of the scientific method; researchers dissect a system and study each piece in isolation. For microorganisms this means separating the organisms and getting them into pure culture. A pure culture is defined as a growth of microorganisms (a culture) that contains one cell type. It is essential in microbiology to be able to obtain and preserve pure cultures.

Over 100 years ago, Robert Koch devised methods to achieve this goal and the methods he developed are essentially still used today. The goals of aseptic technique are two-fold. The first objective is to obtain pure cultures and secondly to prevent cross-contamination. Microorganisms in culture must not escape into the environment and microbes in the environment must not get into the cultures we are studying. It is essential that aseptic technique be understood and practiced correctly. Contaminated cultures are worthless for diagnosis or for doing research on, because it is unclear what microbe is performing any action that is being observed. Aseptic methods commonly used are flame sterilization, tube transfer, streak plates, spread plates and pour plates. Flame sterilization is an easy method to insure sterile transfer of a culture from a source to a growth medium.

Tube transfer is useful for moving inocula from one tube to another. Mechanical dilution by making streak plates is the preferred method for obtaining a pure culture of a microorganism. Finally, spread plates and pour plates are common methods for enumerating microorganisms and are sometimes useful for obtaining isolated colonies.

Flame Sterilization and Tube Transfer

Flame sterilization is a very quick simple method of killing microorganisms on an inoculating loop or needle. The loop or needle is held inside a flame for a few seconds to bring it to redness and then cooled. Once cool, the loop or needle can be used for various culture manipulations. Make sure that the area that contacts the culture is flamed to redness. Also, be patient and let the loop cool down, this usually takes about 15-30 seconds. Learning this technique is essential to everything else we do in microbiology.

Transfer of culture from agar plates to tubes, or from tube to tube, is a common, simple procedure. It is important to perform these transfers in a consistent and rapid manner. The following protocols have been found effective.

To transfer a culture from an agar plate to a broth or agar slant:

- Place the Bunsen burner in front of we and assemble all necessary equipment with in arms reach. Position everything so that we will not burn ourself while trying to inoculate our tubes.
- Label the tube of broth or agar to be inoculated with identifying marks. The culture, the date and our initials for example. Place it in a rack in front of we.
- Holding the inoculating loop handle, flame the entire wire to redness.
- When the wire cools (about 15-30 seconds) remove the lid of the plate with our other hand and obtain an inoculum by removing a small portion of the surface growth on the agar plate. In most cases we will be picking an isolated colony. Choose a well isolated one. Do not dig into the agar. Replace the lid of the plate immediately.
- Hold the tube to be inoculated with the free hand. Remove the cotton plug or cap of the tube with the little finger of the hand holding the needle holder. If a cotton-plugged tube is used, the mouth of the tube should be passed briefly through the flame to singe off dust and lint particles. (Dust or lint may fall into the tube and contaminate the medium.)
- Introduce the inoculum into the tube. When inoculating a tube of broth, rub the wire against the

glass just above the fluid level and then tip the tube slightly to wash the inoculum into the broth. The wire should not be rattled against the sides of the tube to shake an inoculum into the broth; this is unnecessary and may create a dangerous aerosol. If the transfer is made to an agar slant, a single mid-line stroke over its surface is made with the wire or loop.

- Replace the cap or plug.
- Flame the inoculating wire again to redness, slowly to avoid spattering. Put the loop holder down after the wire cools.

Making a Medium

Making medium is as simple as cooking and a crude medium can be made in almost any kitchen with a few utensils and a source of heat. Below is described the production of a chicken broth medium that will grow many common microorganisms. The medium will be sterilized by tyndalization. Simply boiling a medium once, while it will kill most vegetative cells, does not kill endospores.

However, heating encourages the endospores to germinate and a second boiling kills these microbes. The boiling process is repeated a third time to ensure that all spores have germinated and been killed.

- Add 250 ml (about 1 cup) of water into a glass container or some other vessel that can stand boiling water. The container should be something we can cover. Glass bottles that can stand boiling or canning jars work well.
- To this add 15 grams, about 1 tablespoon of instant chicken broth crystals. and stir until dissolved. Cover loosely so that steam can escape, but dust and dirt cannot enter.
- Microwave the broth on high for 3 minutes or until it just begins to boil.
- Let the broth cool and sit for at least 2 hours up to overnight at room temperature. Make sure that the medium vessel is covered to prevent contamination

from the air. At this point the vegetative cells are dead and most endospores will germinate.

- Repeat steps 3 and 4 a second time. This will kill the endospores that germinated on day 1.
- Repeat steps 3 and 4 a third time. This will eliminate the remaining endospores, making the medium sterile.
- Place the medium in a warm place overnight. If our tyndalization was done correctly, our medium should remain clear and free of microbes.
- While we are waiting for our medium, search for a sample we would like to inoculate into it. Almost any sample should work, but natural samples are a good idea for this experiment. Samples coming from food or from our body may contain pathogens. Something we probably want to avoid.
- Check our medium. If it is still clear and sterile, add a pinch of our collected sample to the medium. (The exact amount really does not matter). Again place the medium in a warm place and check it periodically. After a few days, we should start to see the medium become more turbid.
- If we have the equipment, examine a sample of the medium under a microscope. What shapes of microbes are present? Is there more than one species?
- When finished, carefully dispose of the borth. It is basically a spoiled food, but the properties of the spoilage microbe are unknown and should be treated with care.

STREAK PLATES

The streak plate method is a rapid and simple technique of mechanically diluting a relatively large concentration of microorganisms to a small, scattered population of cells. The goal is to obtain isolated colonies on a large part of the agar surface, so that desired species can then be brought into pure culture.

Proper streaking of plates is an indispensable tool in microbiology. In most cases a closed inoculating loop is used for streaking plates. The wire loop should not be badly oxidized or pitted or it will fail to dilute the inoculum and will scratch the surface of the agar. Streak plates can be made from a broth culture, an agar slant or from an agar plate. It is sometimes convenient to suspend a bit of growth from a solid surface in sterile saline and use this as a source of inoculum. Resuspension of colonies or cultures grown on solid surfaces dilutes the culture and makes streak plating easier.

A loopful of inoculum is transferred from the source and put on the agar surface. When using a large inoculum (a turbid culture or growth from a solid surface), a small spot is spread during the initial transfer. If the inoculum is from a lightly turbid suspension, the first phase of the streaking pattern is begun.

The three-phase streaking pattern is recommended for beginners because it is most likely to give satisfactory results with suspensions having a wide range of microbial density.

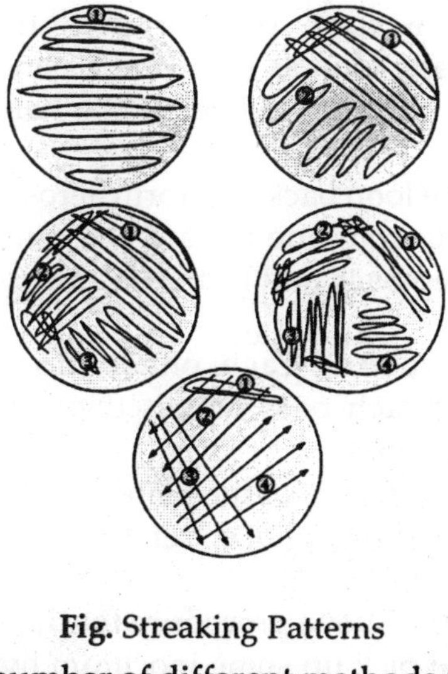

Fig. Streaking Patterns

There are a number of different methods for mechanically diluting microbes on a streak plate. The most common method is spreading microbes across a plate. As the concentration of microbes increases so do the number of phases. Irrespective of the number of phases, loop is flamed between each one. The fifth plate shows an alternative method, where the streaks are not continuous, but are a series of parallel lines. Foobar.

Choosing a streaking pattern is a matter of individual preference and depends upon the number of microorganisms in the sample. Demonstrates the most common patterns, but they are not the only methods.

The object of any streaking pattern is the continuous dilution of the inoculum to give many well isolated colonies. For multi-phase streaking it is crucial to flame the loop before starting the next phase. Note the slight overlap into the previous phase to pick up a small inoculum.

To streak a plate...

- Flame the loop to sterilize it and let cool.
- Position the plate so that the spot of inoculum is nearest the hand not holding the loop (the opposite hand).
- Lift the plate lid with the opposite hand; just enough to get the loop inside and touch the loop to the inoculum spot. It is often helpful to treat the inoculating loop as if it were a pencil - steadying the loop by resting the heel of the hand against the lab bench.
- Move the loop back and forth across the spot and then gradually continue toward the centre of the plate as we sweep back and forth. Use a very gentle and even pressure.
- When creating each phase, do not worry about keeping each pass across the plate separate from previous ones.
- When about 30% of the plate has been covered by the first streaking phase, remove the loop and flame sterilize it.
- Repeat the above procedure for the second phase, but this time pick up some inoculum by crossing into the first phase 2-3 times and then not passing into it again.
- Repeat as necessary for the third and fourth phases. After streaking the plate, flame sterilize the loop before setting it down.

Spread Plates and Dilution Plating

An absolute requirement for a microbiologist is to be able to determine the concentration of microorganisms in a given sample. Various particle-counting devices, spectrophotometric methods and microscopic techniques have been used to count cells. However, one drawback to these methods is that they count dead as well as living cells.

The most common method of enumerating viable cells is the plate-count method. Diluting microorganisms and placing them into petri plates (or plates) for incubation is another essential technique for working with microorganisms. This method suffers from some problems.

First, only those organisms which can grow on the medium and at the temperature and atmospheric conditions of incubation, will divide and develop into colonies. Second, each colony may not represent the progeny from one cell, as two or more cells (those in clusters, chains or otherwise close to one another) can give rise to one colony.

For these reasons the counts obtained from the plate-count method are given as the number of colony-forming units (CFU's) per ml (or gram) rather than the number of cells per ml (or gram). Despite these drawbacks, the plate-count method is a powerful means by which concentrations of viable organisms may be estimated.

Also, if it is desirable to count a specific subgroup of microorganisms in a sample, selective media or special incubation conditions can often be used to encourage the growth of only this class of organism.As microbial quantitation involves the use of pipettes (or micropipettes) in preparing dilutions and inoculating plates, the beginning student in microbiology must become familiar with their use. Due to the possibility of ingesting pathogens and toxic liquids, mouth-pipetting is forbidden in the laboratory! Pipettes are filled and subsequently emptied by the use of propipettes or other pipette bulb.

Pay close attention to the demonstration of their use. Important Safety Consideration: When fitting the pipette and pipette bulb together, use very gentle pressure!! Do not jam these items together! (Force is usually not the answer, a good general rule to live by in this lab and in life for that matter.) The glass pipette will probably break and possibly cause severe injury. Handle the pipette only at the top inch or so. For volumes of 5 ml of less, micropipettes are often the tool of choice.

Instruments are available that are capable of dispensing 5 ml all the way to less than 1 μl (one one-millionth of a liter). Micropipettes have made it possible to miniaturize many experiments and greatly decrease the cost of running them. They are also easy to use and can dispense volumes quickly, increasing the number of experiments that can be performed in a set amount of time. Micropipettes are the tool of choice for small volumes.

Chapter 2

Microbial Cultivation

When microorganisms are cultivated in the laboratory, a growth environment called a medium is used. The medium may be purely chemical (a chemically defined medium), or it may contain organic materials, or it may consist of living organisms such as fertilized eggs.Microorganisms growing in or on such a medium form a culture. A culture is considered a pure culture if only one type of organism is present and a mixed culture if populations of different organisms are present. When first used, the culture medium should be sterile, meaning that no form of life is present before inoculation with the microorganism.

General microbial media. For the cultivation of bacteria, a commonly used medium is nutrient broth, a liquid containing proteins, salts and growth enhancers that will support many bacteria. To solidify the medium, an agent such as agar is added. Agar is a polysaccharide that adds no nutrients to a medium, but merely solidifies it. The medium that results is nutrient agar. Many media for microorganisms are complex, reflecting the growth requirements of the microorganisms.

For instance, most fungi require extra carbohydrate and an acidic environment for optimal growth. The medium employed for these organisms is potato dextrose agar, also known as Sabouraud dextrose agar. For protozoa, liquid media are generally required and for rickettsiae and viruses, living tissue cells must be provided for best cultivation.

For anaerobic microorganisms, the atmosphere must be oxygen free. To eliminate the oxygen, the culture media can be placed within containers where carbon dioxide and hydrogen gas are generated and oxygen is removed from the atmosphere. Commercially available products achieve these conditions. Anaerobic chambers can also be used within closed compartments and technicians can manipulate culture media within these chambers. To encourage carbon dioxide formation, a candle can be burned to use up oxygen and replace it with carbon dioxide.

Special microbial media. Certain microorganisms are cultivated in selective media. These media retard the growth of unwanted organisms while encouraging the growth of the organisms desired. For example, mannitol salt agar is selective for staphylococci because most other bacteria cannot grow in its high-salt environment. Another selective medium is brilliant green agar, a medium that inhibits Gram-positive bacteria while permitting Gram-negative organisms such as *Salmonella* species to grow.

Still other culture media are differential media. These media provide environments in which different bacteria can be distinguished from one another. For instance, violet red bile agar is used to distinguish coliform bacteria such as *Escherichia coli* from noncoliform organisms.

The coliform bacteria appear as bright pink colonies in this media, while noncoliforms appear a light pink or clear. Certain media are both selective and differential. For instance, MacConkey agar differentiates lactose-fermenting bacteria from nonlactose-fermenting bacteria while inhibiting the growth of Gram-positive bacteria. Since lactose-fermenting bacteria are often involved in water pollution, they can be distinguished by adding samples of water to MacConkey agar and waiting for growth to appear.

In some cases, it is necessary to formulate an enriched medium. Such a medium provides specific nutrients that encourage selected species of microorganisms to flourish in a mixed sample. When attempting to isolate *Salmonella* species from fecal samples, for instance, it is helpful to place a sample of the material in an enriched medium to encourage *Salmonella* species to multiply before the isolation techniques begin.

In order to work with microorganisms in the laboratory, it is desirable to obtain them in pure cultures. Pure cultures of bacteria can be obtained by spreading bacteria out and permitting the individual cells to form masses of growth called colonies.

One can then pick a sample from the colony and be assured that it contains only one kind of bacteria. Cultivating these bacteria on a separate medium will yield a pure culture. To preserve microbial cultures, they may be placed in the refrigerator to slow down the metabolism taking place. Two other methods are deep-freezing and freeze-drying.

For deep-freezing, the microorganisms are placed in a liquid and frozen quickly at temperatures below –50°C. Freeze-drying (lyophilization) is performed in an apparatus that uses a vacuum to draw water off after the microbial suspension has been frozen. The culture resembles a powder and the microorganisms can be preserved for long periods in this condition.

STERILIZATION

Wet Heat

Wet heat is the most dependable procedure for the destruction of all forms of microbial life. Steam sterilization generally denotes heating in an autoclave employing saturated steam under a pressure of approximately 15 psi to achieve a chamber temperature of at least 121°C (250°F). The critical factors in insuring the reliability of this sterilization method is:

- Proper temperature and time; and
- The complete replacement of the air with steam (i.e. no entrapment of air).

Some autoclaves utilize a steam activated exhaust valve that remains open during the replacement of air by live steam until the steam triggers the valve to close. Others utilize a pre-cycle vacuum to remove air prior to steam introduction. Physical controls such as pressure gauges and thermometers are widely used but are considered secondary methods of insuring sterilization.

The use of appropriate biological indicators at locations throughout the autoclave is considered the best indicator of sterilization. The biological indicator most widely used for wet heat sterilization is Bacillus stearothermophilus spores.

Dry Heat

Dry heat is less efficient than wet heat sterilization and requires longer times and/or higher temperatures. The specific times and temperatures must be determined for each type of material being sterilized. Generous safety factors are usually added to allow for the variables that can influence the efficiency of this method of sterilization. The moisture of the sterilization environment as well as the moisture history of organisms prior to heat exposure appear to affect the efficiency of dry heat sterilization. Higher temperatures and shorter times may be used for heat resistant materials. The heat transfer properties and the spatial relation or arrangement of articles in the load are critical in insuring effective sterilization.

The advantage of wet heat is a better heat transfer to and into the cell resulting in overall shorter exposure time and lower temperature. Steam sterilization uses pressurized steam at 121-132° C (250-270° F) for 30 or 40 minutes. This type of heat kills all microbial cells including spores, which are normally heat resistant. In order to accomplish the same effect with dry heat in an oven, the temperature needs to be increased to 160-170° C (320-338° F) for periods of 2 to 4 hours.

General Procedures

All materials, equipment, apparatus contaminated with or containing potentially hazardous organisms should be steam sterilized before being washed and stored, or discarded. Autoclaving is the preferred method. Each individual working with biohazardous materials is responsible for sterilization of materials before disposal.

Biohazardous materials should not be placed in autoclaves overnight in anticipation of autoclaving the next day. To minimize hazard to firemen or disaster crews, all biohazardous materials must be placed in an appropriately marked refrigerator or incubator, sterilized, or otherwise confined at the close of each work day.

All autoclaves must be certified for operating efficiency by the periodic use of biological indicator controls and records maintained for three years.

Special precautions should be taken to prevent accidental removal of material from an autoclave before it has been sterilized or the simultaneous opening of both doors on a double door autoclave. Dry hypochlorites, or any other strong oxidizing material, must not be autoclaved with organic materials such as a paper, cloth, or oil:

Oxidizer + Organic material + Heat = Possible Explosion.

All floors, laboratory benches and other surfaces in buildings where biohazardous materials are handled should be disinfected as often as deemed necessary by the supervisor. The surroundings should be disinfected after completion of operations involving plating, pipetting, centrifuging and similar procedures with biohazardous materials. It is the responsibility of the supervisor to determine that the disinfectant and the time and method of exposure is effective against the biological agent(s) used in the facility.

Floor drains should be flooded with water or disinfectant at least once each week in order to fill traps and thus prevent the back flow of sewer gases. Floor cleaning procedures which minimize the generation of aerosols should be used. Wet mopping or wet vacuum pick up is recommended.

Water used to mop floors should contain a disinfectant or disinfectant-detergent. (Dry mopping or dusting should be avoided).

Where wet procedures are not practical, dry vacuum cleaning with a HEPA filter on the exhaust, sweeping compound used with push brooms, or dry dust mop heads treated to suppress aerosolization may be used. Stock solutions of suitable disinfectants will be maintained in each laboratory for disinfection purposes.

Note: working dilutions of certain disinfectants have short shelf lives and must be prepared on a regular basis.

Sterilization Procedures

General criteria for sterilization of typical materials are presented below. Supervisors are encouraged to review the type of materials being handled and to establish standard conditions for sterilization.

Treatment conditions to achieve sterility will vary in relation to the volume of material treated, volume of the autoclave, the contamination level, the moisture content and other factors.

Steam Autoclave:

- *Laundry:* 250°F (121°C) for a minimum of 30 min.
- *Trash:* 250°F (121°C) for at least 45 minutes per bag. Size of the autoclave and of the bags greatly effect sterilization time. Large bags in a small autoclave may require 90 minutes or more.
- *Glassware:* 250°F (121°C) for a minimum of 25 min.
- *Liquids:* 250°F (121°C) for 25 minutes for each gallon.
- *Animals & bedding:* Steam autoclaving not recommended (sterilization time required would be at least 8 hours). Incineration in an approved facility is the recommended treatment of these wastes.

MICROSCOPES

Since microorganisms are invisible to the unaided eye, the essential tool in microbiology is the microscope. One of the first to use a microscope to observe microorganisms was Robert Hooke, the English biologist who observed algae and fungi in the 1660s. In the 1670s, Anton van Leeuwenhoek, a Dutch merchant, constructed a number of simple microscopes and observed details of numerous forms of protozoa, fungi and bacteria.

During the 1700s, microscopes were used to further elaborate on the microbial world and by the late 1800s, the sophisticated light microscopes had been developed. The electron microscope was developed in the 1940s, thus making the viruses and the smallest bacteria (for example, rickettsiae and chlamydiae) visible. Microscopes permit extremely small objects to be seen, objects measured in the metric system in micrometers and nanometers. A micrometer (μm) is equivalent to a millionth of a meter, while a nanometer (nm) is a billionth of a meter. Bacteria, fungi, protozoa and unicellular algae are normally measured in micrometers, while viruses are commonly measured in nanometers. A typical bacterium such as *Escherichia coli* measures about two micrometers in length and about one micrometer in width.

Optical Microscopy

Optical or light microscopy involves passing visible light transmitted through or reflected from the sample through a single or multiple lenses to allow a magnified view of the sample. The resulting image can be detected directly by the eye, imaged on a photographic plate or captured digitally. The single lens with its attachments, or the system of lenses and imaging equipment, along with the appropriate lighting equipment, sample stage and support, makes up the basic light microscope.

Limitations of Optical Microscopy

Limitations of standard optical microscopy (bright field microscopy) lie in three areas;

- The technique can only image dark or strongly refracting objects effectively.
- Diffraction limits resolution to approximately 0.2 micrometre.
- Out of focus light from points outside the focal plane reduces image clarity.

Live cells in particular generally lack sufficient contrast to be studied successfully, internal structures of the cell are colourless and transparent. The most common way to increase contrast is to stain the different structures with selective dyes, but this involves killing and fixing the sample.

Staining may also introduce artifacts, apparent structural details that are caused by the processing of the specimen and are thus not a legitimate feature of the specimen.

These limitations have, to some extent, all been overcome by specific microscopy techniques which can non-invasively increase the contrast of the image. In general, these techniques make use of differences in the refractive index of cell structures. It is comparable to looking through a glass window: we (bright field microscopy) don't see the glass but merely the dirt on the glass.

There is however a difference as glass is a more dense material and this creates a difference in phase of the light passing through. The human eye is not sensitive to this difference in phase but clever optical solutions have been thought out to change this difference in phase into a difference in amplitude (light intensity).

OPTICAL MICROSCOPY TECHNIQUES

Bright Field Optical Microscopy

Bright field microscopy is the simplest of all the light microscopy techniques. Sample illumination is via transmitted white light, i.e. illuminated from below and observed from above. Limitations include low contrast of most biological samples and low apparent resolution due to the blur of out of focus material. The simplicity of the technique and the minimal sample preparation required are significant advantages.

Oblique Illumination and what it Means

The use of oblique (from the side) illumination gives the image a 3-dimensional appearance and can highlight otherwise invisible features. A more recent technique based on this method is Hoffmann's modulation contrast, a system found on inverted microscopes for use in cell culture. Oblique illumination suffers from the same limitations as bright field microscopy (low contrast of many biological samples; low apparent resolution due to out of focus objects), but may highlight otherwise invisible structures.

Dark Field Optical Microscopy

Dark field microscopy is a technique for improving the contrast of unstained, transparent specimens. Darkfield illumination uses a carefully aligned light source to minimise the quantity of directly-transmitted (un-scattered) light entering the image plane, collecting only the light scattered by the sample.

Darkfield can dramatically improve image contrast—especially of transparent objects—while requiring little equipment setup or sample preparation. However, the technique does suffer from low light intensity in final image of many biological samples and continues to be affected by low apparent resolution.

Rheinberg illumination is a special variant of dark field illumination in which transparent, colored filters are inserted just before the condenser so that light rays at high aperture are differently colored than those at low aperture (i.e. the background to the specimen may be blue while the object appears self-luminous yellow). Other colour combinations are possible but their effectiveness is quite variable.

Phase Contrast Optical Microscopy

More sophisticated techniques will show differences in optical density in proportion. Phase contrast is a widely used technique that shows differences in refractive index as difference in contrast. It was developed by the Dutch physicist Frits Zernike in the 1930s The nucleus in a cell for example will show up darkly against the surrounding cytoplasm. Contrast is excellent; however it is not for use with thick objects. Frequently, a halo is formed even around small objects, which obscures detail.

The system consists of a circular annulus in the condenser which produces a cone of light. This cone is superimposed on a similar sized ring within the phase-objective. Every objective has a different size ring, so for every objective another condenser setting has to be chosen.

The ring in the objective has special optical properties: it first of all reduces the direct light in intensity, but more importantly, it creates an artificial phase difference of about a quarter wavelength. As the physical properties of this direct light have changed, interference with the diffracted light occurs, resulting in the phase contrast image.

Differential Interference Contrast Microscopy

Superior and much more expensive is the use of interference contrast. Differences in optical density will show up as differences in relief. A nucleus within a cell will actually show up as a globule in the most often used differential interference contrast system according to Georges Nomarski.

However, it has to be kept in mind that this is an optical effect and the relief does not necessarily resemble the true shape! Contrast is very good and the condenser aperture can be used fully open, thereby reducing the depth of field and maximizing resolution.

The system consists of a special prism in the condenser that splits light in an ordinary and an extraordinary beam. The spatial difference between the two beams is minimal (less than the maximum resolution of the objective). After passage through the specimen, the beams are reunited by a similar prism in the objective.

In a homogeneous specimen, there is no difference between the two beams and no contrast is being generated. However, near a refractive boundary (say a nucleus within the cytoplasm), the difference between the ordinary and the extraordinary beam will generate a relief in the image. Differential interference contrast requires a polarized light source to function; two polarizing filters have to be fitted in the light path, one below the condenser (the polarizer) and the other above the objective (the analyzer).

Note: In cases where the optical design of a microscope produces an appreciable lateral separation of the two beams we have the case of classical interference microscopy, which does not result in relief images, but can nevertheless be used for the quantitative determination of mass-thicknesses of microscopic objects.

FLUORESCENCE MICROSCOPY

When certain compounds are illuminated with high energy light, they then emit light of a different, lower frequency. This effect is known as fluorescence. Often specimens show their own characteristic autofluorescence image, based on their chemical makeup. This method is of critical importance in the modern life sciences, as it can be extremely sensitive, allowing the detection of single molecules.

Many different fluorescent dyes can be used to stain different structures or chemical compounds. One particularly powerful method is the combination of antibodies coupled to a fluorochrome as in immunostaining.

Examples of commonly used fluorochromes are fluorescein or rhodamine. The antibodies can be made tailored specifically for a chemical compound.

For example, one strategy often in use is the artificial production of proteins, based on the genetic code (DNA). These proteins can then be used to immunize rabbits, which then form antibodies which bind to the protein.

The antibodies are then coupled chemically to a fluorochrome and then used to trace the proteins in the cells under study.

Highly-efficient fluorescent proteins such as the green fluorescent protein (GFP) have been developed using the molecular biology technique of gene fusion, a process which links the expression of the fluorescent compound to that of the target protein.

Piston DW, Patterson GH, Lippincott-Schwartz J, Claxton NS, Davidson MW Nikon MicroscopyU: Introduction to Fluorescent Proteins.

Nikon MicroscopyU. Retrieved on 2007-08-22. This combined fluorescent protein is generally non-toxic to the organism and rarely interferes with the function of the protein under study.

Genetically modified cells or organisms directly express the fluorescently-tagged proteins, which enables the study of the function of the original protein in vivo.

Since fluorescence emission differs in wavelength (colour) from the excitation light, a fluorescent image ideally only shows the structure of interest that was labelled with the fluorescent dye.

This high specificity led to the widespread use of fluorescence light microscopy in biomedical research. Different fluorescent dyes can be used to stain different biological structures, which can then be detected simultaneously, while still being specific due to the individual colour of the dye.

To block the excitation light from reaching the observer or the detector, filter sets of high quality are needed. These typically consist of an excitation filter selecting the range of excitation wavelengths, a dichroic mirror and an emission filter blocking the excitation light.

Most fluorescence microscopes are operated in the Epi-illumination mode to further decrease the amount of excitation light entering the detector.

Confocal Laser Scanning Microscopy

Generates the image by a completely different way than the normal visual bright field microscope. It gives slightly higher resolution, but most importantly it provides optical sectioning without disturbing out-of-focus light degrading the image. Therefore it provides sharper images of 3D objects. This is often used in conjunction with fluorescence microscopy.

Deconvolution Microscopy

Fluorescence microscopy is extremely powerful due to its ability to show specifically labelled structures within a complex environment but also because of its inherent ability to provide three dimensional information of biological structures.

Unfortunately this information is blurred by the fact, that upon illumination all fluorescently labeled structures emit light no matter if they are in focus or not.

This means, that an image of a certain structure is always blurred by the contribution of light from structures which are out of focus.

This phenomenon becomes apparent as a loss of contrast especially when using objectives with a high resolving power, typically oil immersion objectives with a high numerical aperture.

Fortunately though, this phenomenon is not caused by random processes such as light scattering but can be relatively well defined by the optical properties of the image formation in the microscope imaging system. If one considers a small fluorescent light source (essentially a bright spot), light coming from this spot spreads out the further out of focus one is.

Under ideal conditions this produces a sort of "hourglass" shape of this point source in the third (axial) dimension. This shape is called the point spread function (PSF) of the microscope imaging system.

Since any fluorescence image is made up of a large number of such small fluorescent light sources the image is said to be "convolved by the point spread function".

Knowing this point spread function means, that it is possible to reverse this process to a certain extent by computer based methods commonly known as deconvolution microscopy. There are various algorithms available for 2D or 3D Deconvolution. They can be roughly classified in non restorative and restorative methods.

While the non restorative methods can improve contrast by removing out of focus light from focal planes, only the restorative methods can actually reassign light to it proper place of origin.

This can be an advantage over other types of 3D microscopy such as confocal microscopy, because light is not thrown away but reused.

For 3D deconvolution one typically provides a series of images derived from different focal planes (called a Z-stack) plus the knowledge of the PSF which can be either derived experimentally or theoretically from knowing all contributing parameters of the microscope.

Sub-diffraction Optical Microscopy Techniques

It is well known that there is a spatial limit to which light can focus: approximately half of the wavelength of the light we are using. But this is not a true barrier, because this diffraction limit is only true in the far-field and localization precision can be increased with many photons and careful analysis (although two objects still cannot be resolved); and like the sound barrier, the diffraction barrier is breakable.

This section explores some approaches to imaging objects smaller than ~250 nm. Most of the following information was gathered (with permission) from a chemistry blog's review of sub-diffraction microscopy techniques.

TYPES OF MICROSCOPES

Various types of microscopes are available for use in the microbiology laboratory. The microscopes have varied applications and modifications that contribute to their usefulness.

The light microscope. The common light microscope used in the laboratory is called a compound microscope because it contains two types of lenses that function to magnify an object.

The lens closest to the eye is called the ocular, while the lens closest to the object is called the objective.

Most microscopes have on their base an apparatus called a condenser, which condenses light rays to a strong beam. A diaphragm located on the condenser controls the amount of light coming through it. Both coarse and fine adjustments are found on the light microscope.

To magnify an object, light is projected through an opening in the stage, where it hits the object and then enters the objective. An image is created and this image becomes an object for the ocular lens, which remagnifies the image. Thus, the total magnification possible with the microscope is the magnification achieved by the objective multiplied by the magnification achieved by the ocular lens.

A compound light microscope often contains four objective lenses: the scanning lens (4X), the low-power lens (10X), the high-power lens (40 X) and the oil-immersion lens (100 X).

With an ocular lens that magnifies 10 times, the total magnifications possible will be 40 X with the scanning lens, 100 X with the low-power lens, 400 X with the high-power lens and 1000 X with the oil-immersion lens.

Most microscopes are parfocal. This term means that the microscope remains in focus when one switches from one objective to the next objective.

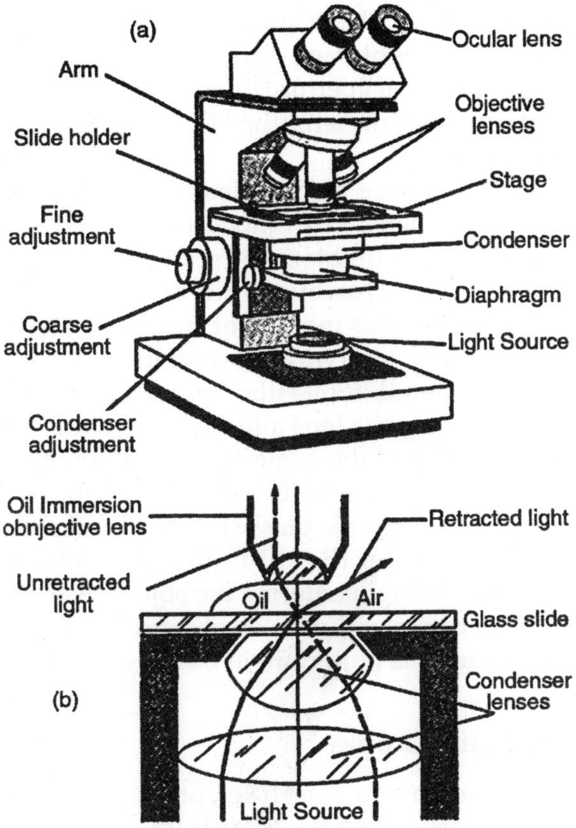

Fig. Light Microscopy.

(a) The Important Parts of a Common Light Microscope.
(b) How Immersion oil Gathers more Light for use in the Microscope

The ability to see clearly two items as separate objects under the microscope is called the resolution of the microscope. The resolution is determined in part by the wavelength of the light used for observing. Visible light has a wavelength of about 550 nm, while ultraviolet light has a wavelength of about 400 nm or less. The resolution of a microscope increases as the wavelength decreases, so ultraviolet light allows one to detect objects not seen with visible light. The resolving power of a lens refers to the size of the smallest object that can be seen with that lens. The resolving power is based on the wavelength of the light used and the numerical aperture of the lens. The numerical aperture (NA) refers to the widest cone of light that can enter the lens; the NA is engraved on the side of the objective lens.

If the user is to see objects clearly, sufficient light must enter the objective lens. With modern microscopes, entry to the objective is not a problem for scanning, low-power and high-power lenses. However, the oil-immersion lens is exceedingly narrow and most light misses it. Therefore, the object is seen poorly and without resolution. To increase the resolution with the oil-immersion lens, a drop of immersion oil is placed between the lens and the glass slide.

Immersion oil has the same light-bending ability (index of refraction) as the glass slide, so it keeps light in a straight line as it passes through the glass slide to the oil and on to the glass of the objective, the oil-immersion lens. With the increased amount of light entering the objective, the resolution of the object increases and one can observe objects as small as bacteria. Resolution is important in other types of microscopy as well.

Other Light Microscopes

In addition to the familiar compound microscope, microbiologists use other types of microscopes for specific purposes. These microscopes permit viewing of objects not otherwise seen with the light microscope. An alternative microscope is the dark-field microscope, which is used to observe live spirochetes, such as those that cause syphilis. This microscope contains a special condenser that scatters light and causes it to reflect off the specimen at an angle. A light object is seen on a dark background.

A second alternative microscope is the phase-contrast microscope. This microscope also contains special condensers that throw light "out of phase" and cause it to pass through the object at different speeds. Live, unstained organisms are seen clearly with this microscope and internal cell parts such as mitochondria, lysosomes and the Golgi body can be seen with this instrument.

The fluorescent microscope uses ultraviolet light as its light source. When ultraviolet light hits an object, it excites the electrons of the object and they give off light in various shades of colour. Since ultraviolet light is used, the resolution of the object increases. A laboratory technique called the fluorescent-antibody technique employs fluorescent dyes and antibodies to help identify unknown bacteria.

Electron Microscopy

The energy source used in the electron microscope is a beam of electrons. Since the beam has an exceptionally short wavelength, it strikes most objects in its path and increases the resolution of the microscope significantly. Viruses and some large molecules can be seen with this instrument. The electrons travel in a vacuum to avoid contact with deflecting air molecules and magnets focus the beam on the object to be viewed. An image is created on a monitor and viewed by the technologist. The more traditional form of electron microscope is the transmission electron microscope (TEM).

To use this instrument, one places ultrathin slices of microorganisms or viruses on a wire grid and then stains them with gold or palladium before viewing. The densely coated parts of the specimen deflect the electron beam and both dark and light areas show up on the image. The scanning electron microscope (SEM) is the more contemporary form electron microscope. Although this microscope gives lower magnifications than the TEM, the SEM permits three-dimensional views of microorganisms and other objects. Whole objects are used and gold or palladium staining is employed.

Staining Techniques

Because microbial cytoplasm is usually transparent, it is necessary to stain microorganisms before they can be viewed with the light microscope. In some cases, staining is unnecessary, for example when microorganisms are very large or when motility is to be studied and a drop of the microorganisms can be placed directly on the slide and observed. A preparation such as this is called a wet mount.

A wet mount can also be prepared by placing a drop of culture on a cover-slip (a glass cover for a slide) and then inverting it over a hollowed-out slide. This procedure is called the hanging drop. In preparation for staining, a small sample of microor ganisms is placed on a slide and permitted to air dry. The smear is heat fixed by quickly passing it over a flame. Heat fixing kills the organisms, makes them adhere to the slide and permits them to accept the stain.

ASimple Stain Techniques

Staining can be performed with basic dyes such as crystal violet or methylene blue, positively charged dyes that are attracted to the negatively charged materials of the microbial cytoplasm. Such a procedure is the simple stain procedure. An alternative is to use a dye such as nigrosin or Congo red, acidic, negatively charged dyes. They are repelled by the negatively charged cytoplasm and gather around the cells, leaving the cells clear and unstained. This technique is called the negative stain technique.

Differential Stain Techniques

The differential stain technique distinguishes two kinds of organisms. An example is the Gram stain technique. This differential technique separates bacteria into two groups, Gram-positive bacteria and Gram-negative bacteria. Crystal violet is first applied, followed by the mordant iodine, which fixes the stain.

Then the slide is washed with alcohol and the Gram-positive bacteria retain the crystal-violet iodine stain; however, the Gram-negative bacteria lose the stain. The Gram-negative bacteria subsequently stain with the safranin dye, the counterstain, used next. These bacteria appear red under the oil-immersion lens, while Gram-positive bacteria appear blue or purple, reflecting the crystal violet retained during the washing step.

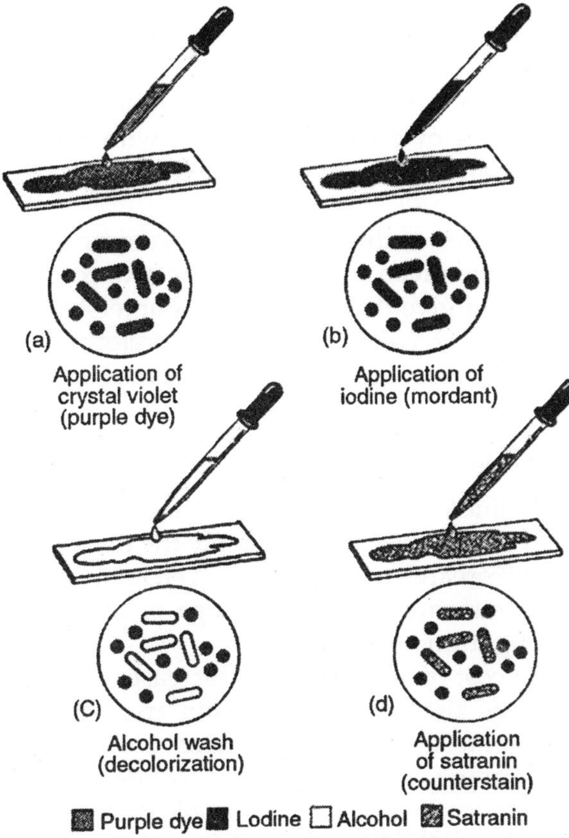

Fig. The Gram Stain Procedure used for Differentiating Bacteria into two Groups

Another differential stain technique is the acid-fast technique. This technique differentiates species of *Mycobacterium* from other bacteria. Heat or a lipid solvent is used to carry the first stain, carbolfuchsin, into the cells. Then the cells are washed with a dilute acid-alcohol solution.

Mycobacterium species resist the effect of the acid-alcohol and retain the carbolfuchsin stain (bright red). Other bacteria lose the stain and take on the subsequent methylene blue stain (blue). Thus, the acid-fast bacteria appear bright red, while the nonacid-fast bacteria appear blue when observed under oil-immersion microscopy. Other stain techniques seek to identify various bacterial structures of importance.

For instance, a special stain technique highlights the flagella of bacteria by coating the flagella with dyes or metals to increase their width. Flagella so stained can then be observed.

A special stain technique is used to examine bacterial spores. Malachite green is used with heat to force the stain into the cells and give them colour. A counterstain, safranin, is then used to give colour to the nonsporeforming bacteria. At the end of the procedure, spores stain green and other cells stain red.

LIGHT MICROSCOPY

The light microscope, so called because it employs visible light to detect small objects, is probably the most well-known and well-used research tool in biology. Yet, many students and teachers are unaware of the full range of features that are available in light microscopes.

Since the cost of an instrument increases with its quality and versatility, the best instruments are, unfortunately, unavailable to most academic programs. However, even the most inexpensive "student" microscopes can provide spectacular views of nature and can enable students to perform some reasonably sophisticated experiments.

A beginner tends to think that the challenge of viewing small objects lies in getting enough magnification. In fact, when it comes to looking at living things the biggest challenges are, in order,

- Obtaining sufficient contrast
- Finding the focal plane
- Obtaining good resolution
- Recognizing the subject when one sees it

The smallest objects that are considered to be living are the bacteria. The smallest bacteria can be observed and cell shape recognized at a mere 100x magnification. They are invisible in bright field microscopes, though. These pages will describe types of optics that are used to obtain contrast, suggestions for finding specimens and focusing on them and advice on using measurement devices with a light microscope.

Types of Light Microscopes

The bright field microscope is best known to students and is most likely to be found in a classroom. Better equipped classrooms and labs may have dark field and/or phase contrast optics. Differential interference contrast, Nomarski, Hoffman modulation contrast and variations produce considerable depth of resolution and a three dimensional effect. Fluorescence and confocal microscopes are specialized instruments, used for research, clinical and industrial applications.

Other than the compound microscope, a simpler instrument for low magnification use may also be found in the laboratory. The stereo microscope, or dissecting microscope usually has a binocular eyepiece tube, a long working distance and a range of magnifications typically from 5x to 35 or 40x. Some instruments supply lenses for higher magnifications, but there is no improvement in resolution. Such "false magnification" is rarely worth the expense.

Bright Field Microscopy

With a conventional bright field microscope, light from an incandescent source is aimed toward a lens beneath the stage called the condenser, through the specimen, through an objective lens and to the eye through a second magnifying lens, the ocular or eyepiece.

We see objects in the light path because natural pigmentation or stains absorb light differentially, or because they are thick enough to absorb a significant amount of light despite being colorless.

A *Paramecium* should show up fairly well in a bright field microscope, although it will not be easy to see cilia or most organelles. Living bacteria won't show up at all unless the viewer hits the focal plane by luck and distorts the image by using maximum contrast. A good quality microscope has a built-in illuminator, adjustable condenser with aperture diaphragm (contrast) control, mechanical stage and binocular eyepiece tube.

The condenser is used to focus light on the specimen through an opening in the stage. After passing through the specimen, the light is displayed to the eye with an apparent field that is much larger than the area illuminated.

The magnification of the image is simply the objective lens magnification (usually stamped on the lens body) times the ocular magnification. Students are usually aware of the use of the coarse and fine focus knobs, used to sharpen the image of the specimen.

They are frequently unaware of adjustments to the condenser that can affect resolution and contrast. Some condensers are fixed in position, others are focusable, so that the quality of light can be adjusted. Usually the best position for a focusable condenser is as close to the stage as possible.

The bright field condenser usually contains an aperture diaphragm, a device that controls the diameter of the light beam coming up through the condenser, so that when the diaphragm is stopped down (nearly closed) the light comes straight up through the centre of the condenser lens and contrast is high.

When the diaphragm is wide open the image is brighter and contrast is low. A disadvantage of having to rely solely on an aperture diaphragm for contrast is that beyond an optimum point the more contrast we produce the more we distort the image. With a small, unstained, unpigmented specimen, we are usually past optimum contrast when we begin to see the image.

Mount the Specimen on the Stage

The cover slip must be up if there is one. High magnification objective lenses can't focus through a thick glass slide; they must be brought close to the specimen, which is why coverslips are so thin.

The stage may be equipped with simple clips (less expensive microscopes), or with some type of slide holder. The slide may require manual positioning, or there may be a mechanical stage (preferred) that allows precise positioning without touching the slide.

Optimize the Lighting

A light source should have a wide dynamic range, to provide high intensity illumination at high magnifications and lower intensities so that the user can view comfortably at low magnifications.

Better microscopes have a built-in illuminator and the best microscopes have controls over light intensity and shape of the light beam. If our microscope requires an external light source, make sure that the light is aimed toward the middle of the condenser. Adjust illumination so that the field is bright without hurting the eyes.

Adjust the Condenser

To adjust and align the microscope, start by reading the manual. If no manual is available, try using these guidelines. If the condenser is focusable, position it with the lens as close to the opening in the stage as we can get it.

If the condenser has selectable options, set it to bright field. Start with the aperture diaphragm stopped down (high contrast). We should see the light that comes up through the specimen change brightness as we move the aperture diaphragm lever.

FOCUS, LOCATE AND CENTRE THE SPECIMEN

Start with the lowest magnification objective lens, to home in on the specimen and/or the part of the specimen we wish to examine. It is rather easy to find and focus on sections of tissues, especially if they are fixed and stained, as with most prepared slides. However it can be very difficult to locate living, minute specimens such as bacteria or unpigmented protists. A suspension of yeast cells makes a good practice specimen for finding difficult objects.

- Use dark field mode (if available) to find unstained specimens. If not, start with high contrast (aperture diaphragm closed down).

- Start with the specimen out of focus so that the stage and objective must be brought closer together. The first surface to come into focus as we bring stage and objective together is the top of the cover slip. With smears, a cover slip is frequently not used.
- If we are having trouble, focus on the edge of the cover slip or an air bubble, or something that we can readily recognize. The top edge of the cover slip comes into focus first, then the bottom, which should be in the same plane as our specimen.
- Once we have found the specimen, adjust contrast and intensity of illumination and move the slide around until we have a good area for viewing.

Adjust Eyepiece Separation, Focus

With a single ocular, there is nothing to do with the eyepiece except to keep it clean. With a binocular microscope (preferred) we need to adjust the eyepiece separation just like we do a pair of binoculars. Binocular vision is much more sensitive to light and detail than monocular vision, so if we have a binocular microscope, take advantage of it.

One or both of the eyepieces may be a telescoping eyepiece, that is, we can focus it. Since very few people have eyes that are perfectly matched, most of us need to focus one eyepiece to match the other image.

Look with the appropriate eye into the fixed eyepiece and focus with the microscope focus knob. Next, look into the adjustable eyepiece (with the other eye of course) and adjust the eyepiece, not the microscope.

Select an Objective lens for Viewing

The lowest power lens is usually 3.5 or 4x and is used primarily for initially finding specimens. We sometimes call it the scanning lens for that reason.

The most frequently used objective lens is the 10x lens, which gives a final magnification of 100x with a 10x ocular lens. For very small protists and for details in prepared slides such as cell organelles or mitotic. Typical high magnification lenses are 40x and 97x or 100x. The latter two magnifications are used exclusively with oil in order to improve resolution.

Move up in magnification by steps. Each time we go to a higher power objective, re-focus and re-centre the specimen. Higher magnification lenses must be physically closer to the specimen itself, which poses the risk of jamming the objective into the specimen.

Be very cautious when focusing. By the way, good quality sets of lenses are parfocal, that is, when we switch magnifications the specimen remains in focus or close to focused. Bigger is not always better. All specimens have three dimensions and unless a specimen is extremely thin we will be unable to focus with a high magnification objective. The higher the magnification, the harder it is to "chase" a moving specimen. The apparent field of an eyepiece is constant regardless of magnification used.

So it follows that when we raise magnification the area of illuminated specimen we see is smaller. Since we are looking at a smaller area, less light reaches the eye and the image darkens. With a low power objective we may have to cut down on illumination intensity. With a high power we need all the light we can get, especially with less expensive microscopes.

Bright field microscopy is best suited to viewing stained or naturally pigmented specimens such as stained prepared slides of tissue sections or living photosynthetic organisms. It is useless for living specimens of bacteria and inferior for non-photosynthetic protists or metazoans, or unstained cell suspensions or tissue sections. Here is a not-so-complete list of specimens that might be observed using bright-field microscopy and appropriate magnifications (preferred final magnifications are emphasized).

- *Prepared slides, stained:* Bacteria (1000x), thick tissue sections (100x, 400x), thin sections with condensed chromosomes or specially stained organelles (1000x), large protists or metazoans (100x).
- *Smears, stained:* Blood (400x, 1000x), negative stained bacteria (400x, 1000x).
- *Living preparations (wet mounts, unstained):* Pond water

(40x, 100x, 400x), living protists or metazoans (40x, 100x, 400x occasionally), algae and other microscopic plant material (40x, 100x, 400x). Smaller specimens will be difficult to observe without distortion, especially if they have no pigmentation.

Care of the Microscope

- Everything on a good quality microscope is unbelievably expensive, so be careful.
- Hold a microscope firmly by the stand, only. Never grab it by the eyepiece holder, for example.
- Hold the plug (not the cable) when unplugging the illuminator.
- Since bulbs are expensive and have a limited life, turn the illuminator off when we are done.
- Always make sure the stage and lenses are clean before putting away the microscope.
- Never use a paper towel, a kimwipe, our shirt, or any material other than good quality lens tissue or a cotton swab (must be 100% natural cotton) to clean an optical surface. We may use an appropriate lens cleaner or distilled water to help remove dried material.
- Organic solvents may separate or damage the lens elements or coatings.
- Cover the instrument with a dust jacket when not in use.
- Focus smoothly; don't try to speed through the focusing process or force anything. For example if we encounter increased resistance when focusing then we 've probably reached a limit and we are going in the wrong direction.

Chapter 3

Microbial Enzymes for Dairy Processing

Indians are known to be lovers of milk and its products. As a sequel to white revolution, India has surged ahead to become the largest milk producer in the world, the production figure touched to 74 million tons for the year 1998. With growing urbanization, demand for processed dairy foods has increased considerably, in particular demand for different cheese varieties and low-lactose milk due to increasing intolerance of human beings to lactose in milk and other milk products.

For improving the quality of milk and milk products, a number of different enzymes from microbial as well as from nonmicrobial sources have potential applications in dairy processing. The use of rennet in cheese manufacture was among the earliest applications of exogenous enzymes in food processing, dating back to approximately 6000 B C. In 1994, the total production of cheese was 8000 metric tons against a total demand of 9000 metric tons.

The projected demand by 2000 A D is around 30,000 metric tons. The use of rennet, as an exogenous enzyme, in cheese manufacture is perhaps the largest single application of enzymes in food processing. In recent years, proteinases have found additional applications in dairy technology, for example in acceleration of cheese ripening, modification of functional properties and preparation of dietic products. Principal among some enzymes that have important and growing applications are lipases and b-galactosidases.

Enzymes with limited applications include glucose oxidase, superoxide dismutase, sulphydryl oxidase, etc. The increasing use of enzymes to produce specific products with characteristic attributes can be emphasized by the world-wide sale of industrial enzymes approximating to US $ 1.6 billion, which is expected to reach US $ 3.0 billion by the year 2008. Almost 45% of this is shared by the food industry and the remaining is shared by detergent (34.4%), textile (11%), leather (2.8%), pulp and paper (1.2%) industries and other industries (5.6%) excluding enzymes for use in diagnostics and therapeutics.

India being the highest producer of milk in the world and consequently the surplus availability of milk in our country has triggered the food and dairy industry to convert the liquid milk into value-added products using biochemical and enzymatic processes.

Microbial Rennets in Dairy Applications

Animal rennet (bovine chymosin) is conventionally used as a milk-clotting agent in dairy industry for the manufacture of quality cheeses with good flavour and texture. Owing to an increase in demand for cheese production world wide – i.e. 4% per annum over the past 20 years, approximating 13.533 million tons – coupled with reduced supply of calf rennet, has therefore led to a search for rennet substitutes, such as microbial rennets. At present, microbial rennet is used for one-third of all the cheese produced world wide.

Rennin acts on the milk protein in two stages, by enzymatic and by nonenzymatic action, resulting in coagulation of milk. In the enzymatic phase, the resultant milk becomes a gel due to the influence of calcium ions and the temperature used in the process. Many microorganisms are known to produce rennet-like proteinases which can substitute the calf rennet. Microorganisms like Rhizomucor pusillus, R. miehei, Endothia parasitica, Aspergillus oryzae and Irpex lactis are used extensively for rennet production in cheese manufacture. Extensive research that has been carried out so far on rennet substitutes has been reviewed by several authors.

Different strains of species of Mucor are often used for the production of microbial rennets.

Whereas best yields of the milk-clotting protease from Rhizomucor pusillus are obtained from semisolid cultures containing 50% wheat bran, R. miehei and Endothia parasitica are well suited for submerged cultivation. Using the former, good yields of milk-clotting protease may be obtained in a medium containing 4% potato starch, 3% soybean meal and 10% barley. During growth, lipase is secreted together with the protease. Therefore, the lipase activity has to be destroyed by reducing the pH, before the preparation can be used as cheese rennet.

Microbial rennets from various microorganisms (marketed under the trade names such as Rennilase, Fromase, Marzyme, Hanilase, etc.) being marketed since the 1970s have proved satisfactory for the production of different kinds of cheese. The molecular and enzymatic properties of chymosins have been studied extensively.

Although the proteolytic specificities of the three commonly used fungal rennets are considerably different from those of calf chymosin, these rennets have been used to produce acceptable cheeses. Recently Novo Nordisk has succeeded in expressing just one proteolytic enzyme from the fungus R. miehei in the well-known organism A. oryzae.

This host organism is able to produce the single protease that cleaves the casein into a glycomacropeptide and para casein by hydrolysing only at the phe105–met106 peptide bond between phenyl alanine and methionine. This monocomponent enzyme product has the trade name Novoren. One major drawback of microbial rennet use in cheese manufacture, is the development of off flavour and bitter taste in the nonripened as well as in the ripened cheeses.

The rennets from microbial sources are more proteolytic in nature in comparison to rennet from animal sources, resulting in production of some bitter peptides during the process of cheese ripening. Hence, attempts have been made to clone the gene for calf chymosin and to express it in selected bacteria, yeasts and molds.

Recombinant Rennets for Cheese Manufacture

Due to shortage of calf stomachs and the economic value of cheese rennet, gene for calf chymosin was one of the first genes for mammalian enzymes that was cloned and expressed in microorganisms. Many different laboratories have cloned the gene for calf prochymosin in Escherichia coli and analysed the structure of the gene as well as the properties of the recombinant chymosin. The expressed proenzyme in E. coli, is present mainly as insoluble inclusion bodies, comprised of reduced prochymosin as well as molecules that are interlinked by disulphide bridges. After disintegration of the cells, inclusion bodies are harvested by centrifugation.

The individual laboratories have reported some differences in the procedure for renaturation of prochymosin from the inclusion bodies, but all have followed the same general scheme. The enzymatic properties of recombinant E. coli chymosin are indistinguishable from those of native calf chymosin. The enzymes were identical when observed by immunodiffusion in gels, but a slight difference was observed by enzyme linked immunosorbent assay (ELISA).

Table. Recombinal Chymosin Proparations at/or Approaching Legal and Commercial Acceptance.

Sourcc of DNA microorganisms	Praducing (brand namc)	Prducing company
Calf abomasums	klayveromyces lactis (Maxiren)	Gist Brocades
Calf abomasums	Aspergillus niger Hansen	Genccor/chr.
(Chyrnoger)		
Synthetic	Escherichic coli	Pfizer (Chymax)

The gene for prochymosin has also been cloned in Saccharomyces cerevisiae; the levels of expression have been reported to be 0.5 to 2.0% of total yeast protein. In yeast, about 20% of the prochymosin can be released in soluble form which can be activated directly; the remaining 80% is still associated with the cell debris.

The cloning of chymosin was carried out without the

prosequence. Though the level of chymosin mRNA was similar to that of prochymosin mRNA, no milk-clotting activity was observed in clones containing the chymosin gene only.

The results suggest that the prosequence is essential for correct folding of the polypeptide chain. The zymogen for the aspartic proteases from R. pusillus, also called mucor rennin, has likewise been cloned and expressed in yeast. Studies on its conversion to active form showed that secretion of R. pusillus protease from recombinant yeast was dependent on glycosylation of the enzyme.

Compared to yeast, the filamentous fungi generally secrete larger quantities of proteins into the culture medium. Furthermore, filamentous fungi secrete the heterologous proteins with correct folding of the polypeptide chain and process correct pairing of sulphydryl groups.

The gene for R. miehei protease has been expressed in A. oryzae. Prochymosin has been expressed in Kluyveromyces lactis, A. nidulans, A. niger and Trichoderma reesei. In most of the cases, the reported yields of the model systems were about 10–40 mg of enzyme per litre of culture medium. However, 3.3 g of enzyme per liter has also been achieved. Yeast, Kluyveromyces lactis, has recently been used as an efficient host for the secretion of recombinant chymosin, which has led to the development of large-scale production process for chymosin. If produced on an industrial scale, the yields will perhaps be of the latter magnitude. Chymosins which are approaching or have marketing and legal acceptance and generally regarded as safe (GRAS).

Most of the companies produce recombinant rennet of cattle calf origin in different microbial hosts, however in India, the major source of milk is buffalo, showing a different composition from that of cow. The rennet from buffalo source has inherent compatibility for clotting buffalo milk.

Therefore, National Dairy Research Institute, Karnal, has taken a lead in cloning gene of buffalo chymosin. The buffalo chymosin cDNA has been cloned in E. coli and partial N-terminal sequencing of the purified buffalo chymosin indicates that it is highly homologous to the cattle chymosin. Several cheese making experiments have been carried out with recombinant chymosin and the general aspects of recombinant chymosin have been dealt with in a report. Since most of the rennet (> 90 %) added to cheese milk is lost in the whey, immobilization would considerably extend its catalytic life. Several rennets have been immobilized, but their efficiency as milk coagulants has been questioned.

So, there is a fairly general support for the view that immobilized enzymes cannot coagulate milk properly owing to inaccessability of the Phe–Met peptide bond of k-casein and that the apparent coagulating activity of immobilized rennets is due to leaching of the enzyme from the support. Different types of conventional cheeses have been successfully made by using recombinant rennet on an experimental or pilot scale.

No major differences have been detected between cheeses made with recombinant chymosins or natural enzymes, regarding cheese yield, texture, smell, taste and ripening. The recombinant chymosins are identical with calf rennet according to the report on biochemical and genetic evidences.

Lactase in Dairy Industry

Lactose, the sugar found in milk and whey and its corresponding hydrolase, lactase or b -galactosidase, have been extensively researched during the past decade. This is because of the enzyme immobilization technique which has given new and interesting possibilities for the utilization of this sugar. Because of intestinal enzyme insufficiency, some individuals and even a population, show lactose intolerance and difficulty in consuming milk and dairy products.

Hence, low-lactose or lactose-free food aid programme is essential for lactose-intolerant people to prevent severe tissue dehydration, diarrhoea and, at times, even death.

Another advantage of lactase-treated milk is the increased sweetness of the resultant milk, thereby avoiding the requirement for addition of sugars in the manufacture of flavoured milk drinks. Manufacturers of ice cream, yoghurt and frozen desserts use lactase to improve scoop and creaminess, sweetness and digestibility and to reduce sandiness due to crystallization of lactose in concentrated preparations. Cheese manufactured from hydrolysed milk ripens more quickly than the cheese manufactured from normal milk.

Technologically, lactose crystallizes easily which sets limits to certain processes in the dairy industry and the use of lactase to overcome this problem has not reached its fullest potential because of the associated high costs. Moreover, the main problem associated with discharging large quantities of cheese whey is that it pollutes the environment. But, the discharged whey could be exploited as an alternate cheap source of lactose for the production of lactic acid by fermentation.

The whey permeate, which is a by-product in the manufacture of whey protein concentrates, by ultrafiltration could be fermented efficiently by Lactobacillus bulgaricus. Lactose can be obtained from various sources like plants, animal organs, bacteria, yeasts (intracellular enzyme), or molds. Some of these sources are used for commercial enzyme preparations.

Lactase preparations from A. niger, A. oryzae and Kluyveromyces lactis are considered safe because these sources already have a history of safe use and have been subjected to numerous safety tests. The most investigated E. coli lactase is not used in food processing because of its cost and toxicity problems.

Properties of Lactase

The properties of the enzyme depend on its source. Temperature and pH optima differ from source to source and with the type of particular commercial preparation. Immobilization of the enzymes, method of immobilization and type of carrier can also influence these optima values.

In general, fungal lactase have pH optima in the acidic range 2.5–4.5 and yeast and bacterial lactases in the neutral region 6–7 and 6.5–7.5, respectively.

The variation in pH optima of lactases makes them suitable for specific applications, for example fungal lactases are used for acid whey hydrolysis, while yeast and bacterial lactases are suitable for milk (pH 6.6) and sweet whey (pH 6.1) hydrolysis. Product inhibition, e.g. inhibition by galactose, is another property which also depends on the source of lactase.

The enzyme from A. niger is more strongly inhibited by galactose than that from A. oryzae. This inhibition can be overcome by hydrolysing lactose at low concentrations by using immobilized enzyme systems or by recovering the enzyme using ultrafiltration after batch hydrolysis. Lactase from Bacillus species are superior with respect to thermostability, pH operation range, product inhibition and sensitivity against high-substrate concentration.

Thermostable enzymes, able to retain their activity at 60°C or above for prolonged periods, have two distinct advantages viz. they give higher conversion rate or shorter residence time for a given conversion rate and the process is less prone to microbial contamination due to higher operating temperature.

Bacillus species have a pH optima of 6.8 and a temperature optima of 65°C. Its high activity for skim milk and less inhibition by galactose has made it suitable for use as a production organism for lactase.

The enzymatic hydrolysis of lactose can be achieved either by free enzymes, usually in batch fermentation process, or by immobilized enzymes or even by immobilized whole cells producing intracellular enzyme.

Although numerous hydrolysis systems have been investigated, only few of them have been scaled up with success and even fewer have been applied at an industrial or semi-industrial level.

Several acid hydrolysis systems have been developed to industry-scale level. Large-scale systems which use free enzyme process have been developed for processing of UHT-milk and processing of whey, using K. lactis lactase (Maxilact, Lactozyme). Several commercial immobilized systems have been developed for commercial exploitation. Snamprogretti process of industrial-scale milk processing technology in Italy is one such working systems.

They make use of fibre-entrapped yeast lactase in a batch process and the milk used is previously sterilized by UHT. For pilot plants, there are three other processes designed and developed to handle milk:

- Gist-Brocades, Rohm GmbH (Germany);
- Sumitomo, Japan. These are continuous processes with short residence times. Processing of whey UF-permeate is accomplished by the system developed by Corning Glass, Connecticut, Lehigh, Valio and Amerace corp. The process by Corning Glass is being applied at commercial scale in the bakers yeast production using hydrolysed whey.

Microbial Enzymes in Accelerated Cheese Ripening

Cheese ripening is a complex process mediated by biochemical and biophysical changes during which a bland curd is developed into a mature cheese with characteristic flavour, texture and aroma. The desirable attributes are produced by the partial and gradual breakdown of carbohydrates, lipids and proteins during ripening, mediated by several agents, viz.

- Residual coagulants,
- Starter bacteria and their enzymes,
- Nonstarter bacteria and their enzymes,
- Indegenous milk enzymes, especially proteinases;
- Secondary inocula with their enzymes.

Proteolysis occurs in all the cheese varieties and is a prerequisite for characteristic flavour development that can be regulated by proper use of the above agents.

Cheese ripening is essentially an enzymatic process which can be accelerated by augmenting activity of the key enzymes. This has the advantage of initiating more specific action for flavour development compared to use of elevated temperatures that can result in accelerating undesirable nonspecific reactions and consequently off flavour development.

Enzymes may be added to develop specific flavours in cheeses, for example lipase addition for the development of

Parmesan or Blue-type cheese flavours. Attempts to accelerate the multiple secondary flavour-forming-reactions, e.g. Strecker degradation, have been scarce. The pathways leading to the formation of flavour compounds are largely unknown and therefore the use of exogenous enzymes to accelerate ripening is mostly an empirical process. Different microbial enzymes used to accelerate cheese ripening.

Studies on the hydrolysis of whole casein and isolated casein components to observe the kinetics and specificity of aspartic proteases in rennin, pepsin and four microbial rennet substitutes indicated the considerable differences in the reaction velocity and the extent of hydrolysis on the rennet curd yield. The rennet enzymes were active even in the later phases of cheese-ripening.

Proteinases and Peptidases

Proteolysis is characteristic of most cheese varieties and is indispensable for good flavour and textural development. Proteinases used in cheese processing include:

- Plasmin,
- Rennet;
- Proteinases (cell wall and/or intracellular) of the starter and nonstarter bacteria.

Approximately 6% of the rennet added to cheese milk remains in the curd after manufacture and contributes significantly to proteolysis during ripening.

Combinations of individual neutral proteinases and microbial peptidases intensified cheese flavour and when used in combinations with microbial rennets reduced the intensity of bitterness caused by the latter.

Acid proteases in isolation cause intense bitterness. Various animal or microbial lipases gave pronounced cheese flavour, low bitterness and strong rancidity, while lipases in combination with proteinases and/or peptidases give good cheese flavour with low levels of bitterness. In a more balanced approach to the acceleration of cheese ripening using mixtures of proteinases and peptidases, attenuated starter cells or cell-free extracts (CFE) are being favoured.

Table. Microbial Enzymes used to Ucoclcrate Chocse Ripening

Enzyme	Source
Microbial lipases	Aspergillus niger A. oryzae Rhizomucor miehei
Lactasce	Streptococcus lactis Ktuyveromyces sp. Escherichia coli
Microbial serline Neutral pmroteinases	A. niger proteinases Bacillus subtilis A. oryzae

PROTEOLYTIC ENZYMES OF LACTIC ACID BACTERIA IN FERMENTED MILK PRODUCTS

The proteolytic system of lactic acid bacteria is essential for their growth in milk and contributes significantly to flavour development in fermented milk products. The proteolytic system is composed of proteinases which initially cleaves the milk protein to peptides; peptidases which cleave the peptides to small peptides and amino acids; and transport system responsible for cellular uptake of small peptides and amino acids. Lactic acid bacteria have a complex proteolytic system capable of converting milk casein to the free amino acids and peptides necessary for their growth. These proteinases include extracellular proteinases, endopeptidases, aminopeptidases, tripeptidases and proline-specific peptidases, which are all serine proteases. Apart from lactic streptococcal proteinases, several other proteinases from nonlactostreptococcal origin have been reported. There are also serine type of proteinases, e.g. proteinases from Lactobacillus acidophillus, L. plantarum, L. delbrueckii sp. bulgaricus, L. lactis and L. helveticus. Aminopeptidases are important for the development of flavour in fermented milk products, since they are capable of releasing single amino acid residues from oligopeptides formed by extracellular proteinase activity.

Other Dairy Enzymes

Other enzymes used for dairy food application include:

- Proteases to reduce allergic properties of cow milk products for infants;
- Lipases for development of lipolytic flavours in speciality cheeses.

The functional properties of milk proteins may be improved by limited proteolysis through the enzymatic modification of milk proteins.

An acid-soluble casein, free of off flavour and suitable for incorporation into beverages and other acid foods, has been prepared by limited proteolysis.

The antigenicity of casein is destroyed by proteolysis and the hydrolysate is suitable for use in milk-protein-based foods for infants allergic to cow milk. Lipolysis makes an important contribution to swiss cheese flavours, due mainly to the lipolytic enzymes of the starter cultures. The characteristic peppery flavour of Blue cheese is due to short-chain fatty acids and methyl ketones. Most of the lipolysis in Blue cheese is catalysed by Penicillium roqueforti lipase, with a lesser contribution from indigenous milk lipase.

The NOVO process for production of enzyme modified cheese (EMC) uses medium-aged cheese which is emulsified, homogenized and pasteurized, after which 'palatase' (a lipase from R. miehei) is added, with or without a proteinase and the blend is ripened at a high temperature for one to four days.

The mixture is reheated, a paste results which is suitable for inclusion in soups, dips, dressings, or snack foods. EMC technology has been developed to produce a range of characteristic cheese flavours and flavour intensities, for example swiss, blue, cheddar, provolo-nemor or romano, suitable for inclusion at low levels in many products.

The claims that exogenous enzymes are effective in accelerating ripening have not led to their wide-spread use, possibly due to their high cost, difficulties in distributing them uniformly in the curd and the possible danger of over-ripening the cheese. The other minor enzymes having limited applications in dairy processing include glucose oxidase, catalase, superoxide dismutase, sulphydryl oxidase, lactoperoxidase and lysozymes.

Glucose oxidase and catalase are often used together in selected foods for preservation. Superoxide dismutase is an antioxidant for foods and generates H_2O_2, but is more effective when catalase is present.

Thermally induced generation of volatile sulphydryl groups is thought to be responsible for the cooked off-flavour in ultra high temperature (UHT) processed milk. Use of sulphydryl oxidase under aseptic conditions can eliminate this defect. The natural inhibitory mechanism in raw milk is due to the presence of low levels of lactoperoxidase (LP), which can be activated by the external addition of traces of H_2O_2 and thiocyanate.

It has been reported that the potential of LP-system and its activation enhances the keeping quality of milk. Cow milk can be provided with protective factors by the addition of lysozyme, making it suitable as an infant milk. Lysozyme acts as a preservative by reducing bacterial counts in milk without affecting the The scope of application of minor enzymes to milk and milk products has been recently reviewed.

The global market for the production of microbial enzymes for use in dairy-products manufacture is considerably large, but is being dominated only by a limited number of enzyme producers. In India the microbial dairy enzymes requirement has been very limited till now.

However, with the advent of technological processes for the manufacture of different varieties of milk products, such as cheeses by the State Dairy Federations, Co-operatives and Private Dairy Product Manufacturers like Amul, Vijaya, Verka, Dynamix, Nestle, Smith Kline Beechem, etc., the markets for the sale of such products in megacities and towns has been slowly growing for the past two to three years.

Presently, many of these microbial enzymes, such as microbial rennets and other enzymes, are being imported. Hence, there is a scope for the production of enzymes such as microbial rennet, lactase, proteinases and lipases indigenously. In the near future, the requirement for these enzymes is bound to increase by leaps and bounds, basically due to requirement of value-added dairy products in the country.

PRODUCTION OF ANTIBIOTICS

The mass production of antibiotics began during World War II with streptomycin and penicillin. Now most antibiotics are produced by staged fermentations in which strains of microorganisms producing high yields are grown under optimum conditions in nutrient media in fermentation tanks holding several thousand gallons.

The mold is strained out of the fermentation broth and then the antibiotic is removed from the broth by filtration, precipitation and other separation methods.

In some cases new antibiotics are laboratory synthesized, while many antibiotics are produced by chemically modifying natural substances; many such derivatives are more effective than the natural substances against infecting organisms or are better absorbed by the body, e.g., some semisynthetic penicillins are effective against bacteria resistant to the parent substance.

Benzylpenicillins and phenoxymethylpenicillins (penicillins G and V, respectively) are produced by fermentation and are the basic precursors of a wide range of semi-synthetic antibiotics, e.g. ampicillin. The amide link may be hydrolysed conventionally but the conditions necessary for its specific hydrolysis, whilst causing no hydrolysis of the intrinsically more labile but pharmacologically essential b-lactam ring, are difficult to attain. Such specific hydrolysis may be simply achieved by use of penicillin amidases (also called penicillin acylases).

Different enzyme preparations are generally used for the hydrolysis of the penicillins G and V, pencillin-V-amidase being much more specific than pencillin-G-amidase. Penicillin amidase may be obtained from E. coli and has been immobilised on a number of supports including cyanogen bromide-activated Sephadex G200.

It represents one of the earliest successful processes involving immobilised enzymes and is generally used in batch or semicontinuous STR processes (40,000 Ukgpenicillin G, 35°C, pH 7.8, 2 h) where it may be reused over 100 times. It has also been used in PBRs, where it has an active life of over

100 days, producing about two tonnes of 6-aminopenicillanic acid $kg^{-1}$of immobolised enzyme.

The penicillin-G-amidases may be used 'in reverse' to synthesise penicillin and cephalosporin antibiotics by non - equilibrium kinetically controlled reactions. Ampicillin has been produced by the use of penicillin-G-amidase immobilised by adsorption to DEAE -cellulose in a packed bed column.

Many other potential and proven antibiotics have been synthesised in this manner, using a variety of synthetic b-lactams and activated carboxylic acids.

PREPARATION OF ACRYLAMIDE

Acrylamide is an important monomer needed for the production of a range of economically useful polymeric materials. It may be produced by the addition of water to acrylonitrile.

$$CH_2 = CHCN + H_2O \rightarrow CH_2{=}CHCONH_2$$

This process may be achieved by the use of a reduced copper catalyst (Cu^+); however, the yield is poor, unwanted polymerisation or conversion to acrylic acid ($CH_2{=}CHCOOH$) may occur at the relatively high temperatures involved (80 – 140°C) and the catalyst is difficult to regenerate. These problems may be overcome by the use of immobilised nitrile hydratase (often erroneously called a nitrilase). The enzyme from Rhodococcus has been used by the Nitto Chemical Industry Co. Ltd, as it contains only very low amidase activity which otherwise would produce unwanted acrylic acid from the acrylamide.

Immobilised nitrile hydratase is simply prepared by entrapping the intact cells in a cross-linked 10% (w/v) polyacrylamide/dimethylaminoe-thylmethacrylate gel and granulating the product. It is used at 10°C and pH 8.0-8.5 in a semibatchwise process, keeping the substrate acrylonitrile concentration below 3% (w/v).

Using 1% (w/v) immobilised-enzyme concentration (about 50,000 U l^{-1}) the process takes about a day. Product concentrations of up to 20% (w/v) acrylamide have been achieved, containing negligible substrate and less than 0.02%

(w/w) acrylic acid. Acrylamide production using this method is about 4000 tonnes per year.

The closely related enzymes cyanidase and cyanide hydratase are used to remove cyanide from industrial waste and in the detoxification of feeds and foodstuffs containing amygdalin.

$$HCN + 2H_2O \rightarrow HCOO^- + NH_4^+$$

$$HCN + H_2O \rightarrow HCONH_2$$

IDENTIFYING USEFUL ANTIBIOTICS

Despite the wide variety of known antibiotics, less than 1% of antimicrobial agents have any medical or commercial value. The most commonly known antibiotic, Penicillin has a highly selective toxicity and therapeutic index (as eukaryotic animal cells do not contain peptidoglycan, they are usually unaffected by it). This is not so for many antibiotics.

Others simply lack advantage over the antibiotics already in use, or have no other practical applications.In order to identify the useful antibiotics, a process of screening is often employed. Using this method, isolates of a large number of microorganisms are cultured and then tested for production of diffusible products which inhibit the growth of test organisms. However, the majority of the resulting antibiotics are already known and must therefore be disregarded.

The remainders must be tested for their selective toxicities and therapeutic activities and the best candidates can be examined and possibly modified.A more modern version of this approach is a rational design programme. This involves screening being directed towards finding new natural products that inhibit specific targets (e.g. a particular step of a metabolic pathway) on microorganisms, rather than tests to show general inhibition of a culture.

Industrial Production Techniques

Antibiotics are produced industrially by a process of fermentation, where the source microorganism is grown in large containers (100,000–150,000 liters or more) containing a liquid growth medium. Oxygen concentration, temperature, pH and nutrient levels must be optimal and are closely monitored and adjusted if necessary. As antibiotics are secondary metabolites, the population size must be controlled very carefully to ensure that maximum yield is obtained before the cells die. Once the process is complete, the antibiotic must be extracted and purified to a crystalline product. This is simpler to achieve if the antibiotic is soluble in organic solvent. Otherwise it must first be removed by ion exchange, adsorption or chemical precipitation.

Strains used for Production

Microorganisms used in fermentation are rarely identical to the wild type. This is because species are often genetically modified to yield the maximum amounts of antibiotics. Mutation is often used and is encouraged by introducing mutagens such as ultraviolet radiation, x-rays or certain chemicals. Selection and further reproduction of the higher yielding strains over many generations can raise yields by 20-fold or more. Another technique used to increase yields is gene amplification, where copies of genes coding for proteins involved in the antibiotic production can be inserted back into a cell, via vectors such as plasmids. This process must be closely linked with retesting of antibiotic production and effectiveness.

TYPES OF ANTIBIOTICS

The great number of diverse antibiotics currently available can be classified in different ways, e.g., by their chemical structure, their microbial origin, or their mode of action. They are also frequently designated by their effective range. Tetracyclines, the most widely used broad-spectrum antibiotics, are effective against both Gram-positive and Gram-negative bacteria, as well as against rickettsias and psittacosis-causing organisms. Ciprofloxacin (Cipro) is another broad-spectrum antibiotic, effective in the treatment of mild infections of the urinary tract and sinuses.

The medium-spectrum antibiotics bacitracin, the erythromycins, penicillin and the cephalosporins are effective primarily against Gram-positive bacteria, although the streptomycin group is effective against some Gram-negative and Gram-positive bacteria. Polymixins are narrow-spectrum antibiotics effective against only a few species of bacteria.

Administration and Side Effects

Antibiotics are either injected, given orally, or applied to the skin in ointment form. Many, while potent anti-infective agents, also cause toxic side effects. Some, like penicillin, are highly allergenic and can cause skin rashes, shock and other manifestations of allergic sensitivity. Others, such as the tetracyclines, cause major changes in the intestinal bacterial population and can result in superinfection by fungi and other microorganisms. Chloramphenicol, which is now restricted in use, produces severe blood diseases and use of streptomycin can result in ear and kidney damage. Many antibiotics are less effective than formerly because antibiotic-resistant strains of microorganisms have emerged.

Nonmedical Use

Antibiotics have found wide nonmedical use. Some are used in animal husbandry, along with vitamin B_{12}, to enhance the weight gain of livestock. However, some authorities believe the addition of antibiotics to animal feeds is dangerous because continuous low exposure to the antibiotic can sensitize humans to the drug and make them unable to take the substance later for the treatment of infection.

In addition low levels of antibiotics in animal feed encourage the emergence of antibiotic-resistant strains of microorganisms. Drug resistance has been shown to be carried by a genetic particle transmissible from one strain of microorganism to another and the presence of low levels of antibiotics can actually cause an increase in the number of such particles in the bacterial population and increase the probability that such particles will be transferred to pathogenic, or disease-causing, strains. Antibiotics have also been used to treat plant diseases such as bacteria-caused infections in tomatoes, potatoes and fruit trees. The substances are also used in experimental research.

Chapter 4

Concept of Fermentation

When our early ancestors made alcoholic beverages, they used a bioprocess. The combination of yeast cells and nutrients (cereal grains) formed a fermentation system in which the organisms consumed the nutrients for their own growth and produced by-products (alcohol and carbon dioxide gas) that helped to make the beverage. Bioprocess technology is an extension of ancient techniques for developing useful products by taking advantage of natural biological activities such as production of enzymes (used, for example, in food processing and waste management) and antibiotics. Use of living material offers several advantages over conventional chemical methods of production.

They usually require lower temperature, pressure, and pH. They can use renewable resources as raw materials and greater quantities can be produced with less energy consumption. With the development of automated and computerized equipment, it is becoming much easier to accurately monitor reaction conditions and thus increase production efficiency.

As advances in bioprocess technology, particularly separation and purification techniques, are made, commercial firms will be able to economically produce these substances in large amounts, and thus make them available for use in medical research, food processing, agriculture, pharmaceutical development, waste management, and numerous other fields of science and industry.

Knowledge of the metabolism of microbial cells is indispensable for:

- The optimization of bioreactor operation,
- The selection of the best microorganism for a given biochemical conversion,
- Targeted or evolutionary improvement of an existing microorganism
- Design (metabolic engineering) of microorganisms with new reactivities and/or morphology. This helps improve the efficiency of bioprocesses by focusing on microbial performance. Growth and production characteristics of the microorganisms or cell culture must be measured as a function of environment. Shake flasks of 250 ml to 1 liter capacity are used for these experiments. Varying conditions for optimal growth of microorganisms and their productivity are medium composition, pH, temperature and other environmental conditions. How is the working of the bench top fermenter which is also a bioreactor works?

Bench Top Fermenter (bioreactor) This is the first step of the scale up process for large-scale operation of the bioprocess for which optimum conditions for the growth are established in the shake flask experiments above. The bench top fermenter usually is of 1 to 2 liter capacity. It is equipped with instruments for measuring and adjusting:

- Temperature
- pH
- Dissolve O_2
- Concentration of substrate
- Concentration of product
- Agitator speed
- Other process variables

Process can be more closely controlled than in Shake flask experiments. Information is collected on:

- O_2 requirements of the cells
- Sensitivity of the cells to shear
- Foaming characteristics

Many proteins and polymers produced will cause extensive foaming. The main concern of this study is whether optimal conditions can be provided or not for the cell growth in a reactor and their determination. Information gathered from this study is used for the further scale-up of the bioreactor.

The information obtained includes:

- Mass transfer coefficients (concentration gradients)
- Mixing time (a parameter used for determining mixing efficiency)
- Gas hold up $\varepsilon = \frac{v_g}{v_g + v_2}$
- Vg is the volume of gas bubbles in the reactor
- V_2 is the volume of liquid
- Power number $\frac{P}{rN^3D^3}$
- P is the power required r is the viscosity D_1^5

Impeller shear rate: It is also decided whether the bioreactor must be operated for batch process or continuous process. At this stage, the commercial viability of the process is also evaluated.

FERMENTATION

Fermentation is the process of deriving energy from the oxidation of organic compounds, such as carbohydrates, using an endogenous electron acceptor, which is usually an organic compound.

Sugars are the common substrate of fermentation, and typical examples of fermentation products are ethanol, lactic acid, and hydrogen. However, more exotic compounds can be produced by fermentation, such as butyric acid and acetone. Yeast carries out fermentation in the production of ethanol in beers, wines and other alcoholic drinks, along with the production of large quantities of carbon dioxide.

Fermentation occurs in mammalian muscle during periods of intense exercise where oxygen supply becomes limited. Decom-position of foodstuffs generally accompanied by the evolution of gas. The best-known example is alcoholic fermentation, in which sugar is converted into alcohol and carbon dioxide. During fermentation organic matter is decomposed in the absence of air (oxygen); hence, there is always an accumulation of reduction products, or incomplete oxidation products. Some of these products are of importance to humans, and fermentation has therefore been used for their manufacture on large industrial scale.

There are also many microbiological processes that go on in the presence of air while yielding incomplete oxidation products. Good examples are the formation of acetic acid (vinegar) from alcohol by vinegar bacteria, and of citric acid from sugar by certain molds (for example, *Aspergillus niger*). These microbial processes, too, have gained industrial importance, and are often referred to as fermentations, even though they do not conform to L. Pasteur's concept of fermentation as a decomposition in the absence of air.

Anaerobic metabolism. Used generally of alcohol fermentation of sugars, also production of lactic acid, citric acid, etc., by micro-organisms, which may be yeasts, bacteria or fungi.

The breakdown of organic substances by organisms to release energy in the absence of oxygen. It is especially applied to the anaerobic breakdown of carbohydrates by yeasts to produce alcohol and carbon dioxide, and the bacterial breakdown of milk sugar to give lactic acid (as in the production of cheese and yoghurt).

Process by which the living cell is able to obtain energy through the breakdown of glucose and other simple sugar molecules without requiring oxygen. Fermentation is achieved by somewhat different chemical sequences in different species of organisms. Two closely related paths of fermentation predominate for glucose. When muscle tissue receives sufficient oxygen supply, it fully metabolizes its fuel glucose to water and carbon dioxide.

However, at times of strenuous activity, muscle tissue uses oxygen faster than the blood can supply it. During this anaerobic condition, the six-carbon glucose molecule is only partly broken down to two molecules of the three-carbon sugar called lactic acid.

This process, called lactic acid fermentation, also occurs in many microorganisms and in the cells of most higher animals. In alcoholic fermentation, such as occurs in brewer's yeast and some bacteria, the production of lactic acid is bypassed, and the glucose molecule is degraded to two molecules of the two-carbon alcohol, ethanol, and to two molecules of carbon dioxide.

Many of the enzymes of lactic acid and alcoholic fermentation are identical to the enzymes that bring about the metabolic conversion known as glycolysis. Alcoholic fermentation is a process that was known to antiquity. Before 2000 B.C. the Egyptians apparently knew that crushed fruits stored in a warm place would produce a substance with a pleasant intoxicating power.

By 1500 B.C. the production of beer from germinating cereals (malt) and the preparation of wines from crushed grapes were established arts in most of the Middle East. Aristotle believed that grape juice was an infantile form of wine and that fermentation was, therefore, the maturation of the grape extract.

Interest in the process of fermentation has continued through the ages, and much of modern biochemistry, especially enzyme studies, has emerged directly from early studies on the fermentation process. One of the earliest laboratories established for the study of biological chemistry was that founded in Copenhagen in 1875 and financed by the brewing family of Jacob Christian Jacobsen.

Fermentation is one of the oldest known food preservation techniques. Along with drying and salting, fermentation was a key method of extending the life of foods, allowing them to be available, and eaten safely, in times of scarcity or seasonal nonavailability. These methods helped allow the transition from hunting and gathering to organized food cultivation and storage, which took place some ten to fifteen thousand years ago in the Middle East.

Fermentation involves the action of desirable microorganisms, or their enzymes, on food ingredients to make biochemical changes, which cause significant modification to the food. Often lactic-acid bacteria convert the carbohydrate energy source of food, such as lactose in milk, to lactic acid; examples are yogurt and cheeses from milk, and pickles from fruits and vegetables.

Alternatively, yeasts, often of the *Saccharomyces* species, may convert the glucose to ethanol and carbon dioxide in leavened breads, or the sugars in grain or fruit beverages to beers and wines. Molds also can be active in certain fermentations, such as Stilton cheese and soy sauce. It is estimated that about one-third of all the food we consume is fermented. World estimates for beer consumption are about 22 million gallons, and a total of 15 million tons of some one thousand varieties of cheese are eaten annually.

FERMENTED BEVERAGES AND FOODS

Fermentation is often the key to the safe, enjoyable consumption of perishable food materials, as it changes their composition, flavour, and texture. For example, milk is a nutritious but highly perishable beverage. Originally, in the Middle East, milk carried in animal-skin containers, often on horseback, would sour naturally, to produce acidic fermented milk.

The combined action of the two lactic-acid bacteria, *Streptococcus lactis*, producing lactic acid, and *Lactobacillus bulgaricus*, producing lactic acid and acetaldehyde, a major contributor to flavour, are involved in yogurt production. The Tartars of Central Asia used the milk of horses, donkeys, or camels to produce a fizzy, gray acidic and alcoholic drink, *kumiss*, in which yeasts were active.

In acid conditions, the milk protein, casein, denatures and is precipitated to form a curd, producing cottage and soft cheese. By stirring and pressing, whey is removed and a more solid curd is produced, which by ripening or maturation produces semi-hard or hard cheeses. Surface-active bacteria of *Brevibacterium linens* are active in producing the aroma of Limburger type cheeses, while the blue molds of the genus *Penicillium* give Stilton and Gorgonzola cheeses their character. The use of *Saccharomyces* yeasts has allowed the production of a range of fermented beverages, enabling safe consumption of liquid when fresh water supplies are not available.

Lagers, the light golden, gassy beverage made by "bottom" yeast fermentation of cereal extracts, were first made in the regions of Germany and Czechoslovakia, but are now produced and consumed throughout the world. In Africa, a thick, sour alcoholic beverage is made from sorghum or millet, or sometimes maize or banana.

These sorghum beers are important sources of nutrients, particularly B vitamins, to people on marginal diets in these regions. The Romans planted extensive vineyards in North Africa to harvest and ferment their grapes into wine, thereby producing a fermented beverage that could be readily stored, transported, and consumed when and where required.

Distillation of these alcoholic beverages, such as whiskey from beers, brandies from grape wines, or arrack from palm or rice wine, further extend our range of drinks and play important cultural roles in festivities.

FERMENTATION VESSELS AND STARTER CULTURES

Art meets science in the production of fermented foods. Traditional practices are passed down through generations of producers, often small in scale, and consumption patterns often have great cultural importance. In Scandinavia, traditionally the brides and mothers jealously guard their own supplies of sourdough starters, so that they can always make the desired bread for their partners and families.

In West Africa, a homeowner keeps a supply of *dawadawa,* a dried fermented African locust bean paste (*Parkia* species); it is used to give everyday soups and stews the desired "meaty" flavour, while also providing important nutrients, such as riboflavin, the B vitamin that protects against blindness, which is endemic to the region due to nutritional deficiency. In Korea, few meals are complete without *kimchi,* a pickled fermented cabbage, which may also contain fish and other components. The practice of every home having their own *kimchi* jars, often on their verandahs, originated as a way of preserving vegetables through the cold winter season, providing year-round vitamin C. *Kimchi* together with *kochujang,* the fermented red pepper paste, give Korean preparations a unique and characteristic attractive colour and flavour.

Where food fermentation occurred naturally as conditions favored particular organisms, an important art arose to encourage the desired fermentation organisms, while preventing undesirable microorganisms from developing, for successful fermented food production.

Food storage often took place in earthenware vessels, whose semipermeable inner walls were difficult to clean completely. This allowed a biofilm of desirable microorganisms to remain, to initiate a successful fermentation of the next batch of food. Because of their significance, the vessels themselves were artistically designed and treasured. Interesting examples can be seen in museum collections, such as the Nezu Museum in Tokyo, Japan, and a museum dedicated to *kimchi* in Seoul, South Korea.

In Europe, the fermented meat producers, while using ceramic or metallic vats with smoother, more easily cleaned surfaces, developed the technique of "backslopping" to introduce a small quantity of the fermenting liquor from the previous batch of meat to initiate successful fermentation.

In many cases, dried grains or balls of the derived fermenting microorganisms on cereal or other substrates would be used to start fermentation. Baker's yeast may be used in this work. Kefir grains are used in North Africa, the Middle East, and Russia for production of kefir, *laban,* or *leben* fermented milks. *Ragi* is used in Indonesia and throughout East and Southeast Asia as *inoculum* for *lao-chao* and other fermented foods.

CULTURAL DIVERSITY

The production, consumption, and enjoyment of different fermented foods reflects the diversity of cultures and cuisines that make up our varied world. In Chinese and Japanese cuisines, *shoyu,* or soy sauce, is added almost universally to dishes, while the Indian vegetarian diet depends on fermented cereals and legumes, often in combinations, as in *dosas* and *vadas*. The art and science of fermenting meat to a wide range of salamis are vital to the enjoyment of Eastern and Central Europeans, while Italian food market stall holders proudly display their mold-covered fermented sausages and traditional cheeses.

As people migrate, they normally carry their traditional fermented food practices with them. The range of fermented cheeses and meats in Latin America reflects the European origins of these populations, and the wineries of Chile were originally established by French families. Consumers of imported wine, chocolate, coffee, or tea are all beneficiaries of the internationalism and significance of fermented foods.

MICROBIAL GROWTH KINETICS

Microbial growth can be defined as an orderly increase in cellular components, resulting in cell enlargement and eventually leading to cell division. This definition is not strictly accurate as it implies that a consequence of growth is always an increase in cell numbers.

However, under certain conditions growth can occur without cell division, for example, when cells are synthesizing storage compounds, e.g. glycogen or poly-b-hydroxybutyrate. In this situation the cell numbers remain constant, but the concentration of biomass continues to increase. This is also true for coenocytic organisms, such as some fungi, that are not divided into separate cells. Their growth results only in increased Size.

BACTERIAL GROWTH KINETICS

Growth kinetics of homogeneous unicellular suspension. cultures can be modeled using differential equations in a continuum model. However, filamentous growth and growth in heterogeneous cell aggregates and assemblage, particularly biofilms, colonies, flocs, mats and pellicles, is much more complex. In fact, heterogeneous systems require a very different approach using cellular automaton and Swarm system models, e.g. BacSim. The growth kinetics of filamentous organisms and heterogeneous systems will not be discussed here.

MODELING THE MICROBIAL GROWTH KINETICS

The growth model that will be examined is bacterial binary fission in homogeneous suspension cultures, where cell division produces identical daughter cells. Each time a cell divides is called a generation and the time taken for the cell to divide is referred to as the generation time. Therefore, the generation time or doubling time *(td)* is the time required for a microbial population to double.

Theoretically, after one generation, both the microbial cell population and biomass concentration have doubled. However, as previously stated, under certain conditions growth can be associated with an increase in biomass and not cell numbers.

Also, the generation time recorded during microbial growth is in reality an average value, as the cells will not be dividing at exactly the same rate. At anyone time there are cells at different stages of their cell cycle. This is termed asynchronous growth. However, under certain conditions synchronous growth can be induced so that all cells divide simultaneously, which is a useful research tool in the study of microbial physiology.

Microbial fermentations in liquid media can be carried out under different operating conditions, i.e. batch growth, fed-batch growth or continuous growth.

Batch growth involves a closed system where all nutrients are present at the start of the fermentation within a fixed volume.

The only further additions may be acids or bases for pH control, or gases (e.g. aeration, if required). In fed-batch systems fresh medium or medium components are fed continuously, intermittently or are added as a single supplement and the volume of the batch increases with time. Continuous fermentations are open systems where fresh medium is continuously fed into the fermentation vessel, but the volume remains constant as spent medium and cells are removed at the same rate.

BATCH GROWTH

During batch fermentations the population of microorganisms goes through several distinct growth phases: lag, acceleration, exponential growth, deceleration, stationary and death. In the lag phase virtually no growth occurs and the microbial population remains relatively constant. Nevertheless, it is a peripd of intense metabolic activity as the microbial inoculum adapts to the new environment. When cells are inoculated into fresh medium they may be deficient in essential enzymes, vitamins or cofactors, etc., that must be synthesized in order to utilize available nutrients, prior to cell division taking place.

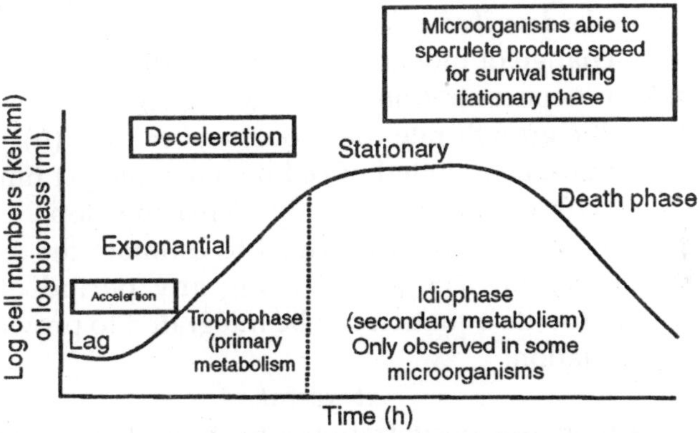

The chemical composition of the fermentation media influences the length of the lag phase. It is usually longer if the inoculum was grown up using a carbon source different from that in the fresh medium, because the cells must synthesize enzymes required to catabolize the new substrate. Physiological stress may also have an effect, especially as cells are often transferred from an inoculum medium of low osmotic pressure (low solute concentration) to fresh medium of higher osmotic pressure (high solute concentration). Other factors influencing the length of the lag phase are the age, concentration, viability and morphology of the inoculum. Generally, inocula prepared from cells harvested in the exponential growth phase (period of most rapid growth) exhibit shorter lag phases than those harvested from subsequent stages.

Once the cells have adapted to their new environment they enter the acceleration phase. Cell division occurs with increasing frequency until the maximum growth rate (mmax) for the specific conditions of the batch fermentation is reached. At this point exponential growth begins and cell numbers/ biomass increase at a constant rate.

Mathematically, this exponential growth can be described by two methods; one is related to biomass *(x)* and the other to cell numbers *(N)*. For cell biomass, growth can be considered as an autocatalytic reaction. Therefore, the rate of growth is dependent on the biomass concentration, i.e. catalyst, that is present at any given time.

This can be described as follows:

rate of change of biomass is $dx/dt = \mu \times$ (5.1)

where x = concentration of biomass (g/L),

μ= specific growth rate (per hour) and t= time (h).

When a graph is plotted of cell biomass against time, the product is a curve with a constantly increasing slope.

Equation can also be rearranged to estimate the specific growth rate (μ): m= *1/x dxldt* During any period of true exponential growth, equation can be integrated to provide the following equation:

$$xt=xoemt$$

where xt = biomass concentration after time' t, xo = biomass concentration at the start exponential growth, and e = base of the natural logarithm.

Taking natural logarithms, loge (In), gives: In Xt= In Xo + μt: This equation is of the form y = c (intercept on y axis) + mx where m = gradient, which is the general equation for a straight-line graph. For cells in exponential phase, a plot of natural log of biomass concentration against time, a semilog plot, should yield a straight line with the slope (gradient) equal to μ, or m= (In xt-In $xo)/t$ (note: when plotting log10 values instead of the natural log, the gradient of the semilog plot is equal to m/2.303;).

A second approach is to examine growth in relation to cell number, where the number of cells at the start of exponential growth is No. If we consider a hypothetical case where the microbial cell population at this point of exponential growth is one (No=1), we can describe binary fission as follows:

No. of divisions 0 1, 2,3... *n*

No. of cells 1, 2, 4, 8...2*n*

Consequently, after a period of exponential growth, time *(t)*, the number of cells *(Nt)* is given by,

$$Nt = No\ 2n$$

where n=the number of divisions, No = initial cell number.

Taking natural logarithms gives: In *Nt* = lnNo+ *n*ln2: Therefore, the number of divisions *(n)* that have taken place is given by In *Nt*-In No n = In2: The number of divisions per unit time during this period of exponential growth is determined by dividing by the time period *(t)*:

$$n = lnNt\text{-}lnNo$$

$$tln2$$

where *n*/*t* = division rate constant (average number of generations per hour).

Often, we are not really interested in the number of divisions per hour, unit time, but rather in the mean generation time or doubling time *(td)*, that is, the time required to undergo a single generation that doubles the population.

Thus,

$$td = t = t\ ln2\ n\ In\ Nt\text{-}In\ No$$

During exponential growth, when all nutrients are supplied in excess and are therefore non-limiting, there is a direct relationship between cell numbers and biomass concentration, assuming that mean cell size is constant. This is balanced growth, and a direct relationship between specific growth rate and doubling time can also be established. However, under conditions where an essential nutrient becomes limiting, unbalanced growth arises, and variations in cell numbers *(N)* and biomass *(x)* concentration occur, as during the synthesis of cell storage compounds.

If we consider a situation where at time zero, the cell biomass is *xo*, then after a fixed period of time *(t)* of exponential growth, equivalent to one doubling time *(td)*, the microbial biomass will double to *2xo*, i.e. *xt=2xo*, when *t=td*, Substituting these parameters into equation gives *2xo* = *xoe* mt d

Taking natural logarithms produces:

$$\text{In}2xo = \text{In } xo + \text{mtd}$$

or

$$\text{mtd} = \text{In } 2.$$

Therefore, in this case,

$$td = \frac{0.693}{\mu}$$

Equation, rate of change of biomass *(dx/dt*=m*x)*, predicts that growth will occur indefinitely. However, during batch growth the microorganisms are continuously metabolizing the finite supply of nutrients available in the fermentation broth. After a certain time the growth rate decreases and eventually stops. This cessation of growth can be due to depletion of essential nutrients (carbon source, essential amino acids, etc.) or the build-up of toxic metabolites, such as ethanol and lactic acid, or a combination of nutrient depletion and toxin accumulation. Monod showed that growth rate is an approximate hyperbolic function of the concentration of the growth-limiting nutrient(s). This impact of essential nutrient depletion on-growth can be described mathematically by the Monod equation, in a form similar to that used in biochemistry, where Michaelis-Menten kinetics define the rate of an enzymecatalysed reaction in relation to its substrate concentration.

$$\mu = \frac{\mu_{max} S}{Ks + S}$$

where μ_{max} = maximum specific growth (per hour) of the cells, i.e. when substrate concentration is not limiting; S = concentration of limiting substrate (g/L); *Ks* = saturation constant, concentration (g/L) of limiting nutrient enabling growth at half the maximum specific growth rate, i.e. *m*.=1/2 mmax and is a measure of the affinity of the cells for this nutrient.

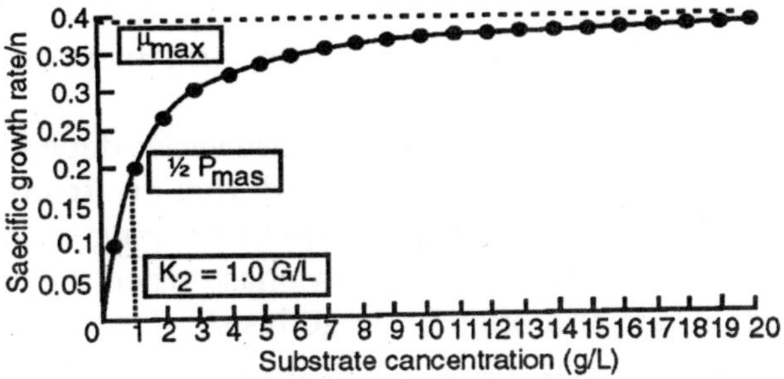

Fig. The Relationship Between Substrate Concentration

and specific growth rate.

When a microorganism is provided with the limiting substrate at a concentration much greater than the *Ks,* and with all other nutrients in excess, the microorganism will grow exponentially at its maximum rate, i.e. when S» *Ks,* then $\mu.=\mu._{max}$.

However, as the level of this substrate decreases, it eventually becomes limiting and can no longer sustain μ_{max}. This is the beginning of the deceleration phase. As the residual concentration of the limiting substrate approaches *Ks* and then falls below this concentration, there is an accompanying gradual decrease in growth rate (μ). The growth rate of a microorganism with a very high affinity for a rate-limiting substrate (i.e. a low *Ks)* will not be affected until the substrate concentration becomes very low.

However, where there is a low affinity for the limiting substrate (i.e. a high *Ks),* the growth rate will begin to fall even at relatively high substrate concentrations and the organism exhibits a longer deceleration phase. The specific growth rate of the microorganism continues decelerating until all of the available limiting substrate is metabolized. Growth is no longer sustainable and the cells enter the stationary phase. At this point, the overall growth rate has declined to zero and there is no net change in cell numbers/biomass (rate of cell division equals rate of cell death). However, the microorganisms are still metabolically active, involved in metabolizing intracellular storage compounds, utilizing nutrients released from lysed cells, and in some cases producing secondary metabolites.

The duration of the stationary phase varies depending on the microorganisms involved and the environmental conditions. For cells unable to survive by forming spores, this is followed by an exponential death phase when the cells die at a constant rate and often undergo lysis.

APPLICATION OF BATCH FERMENTATIONS

During batch fermentations certain environmental conditions continually change, particularly nutrient and product concentrations, as does the specific growth rate, because the cells must pass through the sequence of growth phases described above.

Consequently, the system never achieves steady-state conditions. A further disadvantage is that several distinct practical stages are associated with the operation of a batch fermentation:

- Charging of the fermenter with fresh medium;
- Sterilization of the fermenter and medium;
- Inoculation of the fermenter;
- Production of microbial products;
- Harvesting of biomass and spent fermentation broth; and finally
- Cleaning of the vessel.

This has major economic implications for industrial processes. For a considerable period of time, the fermentation vessel is not producing microbial products, but is being cleaned, filled, sterilized, etc. The non-productive period is referred to as the down-time of the fermenter.

OPTIMIZATION OF BATCH FERMENTATIONS

During the development of a batch process, key growth parameters can be determined that enable the production of a given microbial product to be optimized, whether it is the biomass itself or a specific metabolite. An important parameter is the yield coefficient (Y), which is determined on the basis of the quantity of rare limiting nutrient, normally the carbohydrate source, converted into the microbial product. In respect of biomass, it is defined as:

$$x = Y\,x/s(S\text{-Sr})$$

where x = biomass concentration (g/L), $Y\,x/s$ =*yield* coefficient (g biomass/g substrate utilized), S = initial substrate concentration (g/L), and Sr = residual substrate concentration (g/L).

In the case of biomass production, the yield coefficient relates to the quantity of biomass produced per gram of substrate utilized. Therefore, the higher the yield coefficient, the greater the percentage of the original substrate converted into microbial biomass.

For microbial metabolic products *(p)* the yield coefficient is related to the quantity of metabolite produced in relation to the quantity of substrate used *Y p/s)'* Determination of yield coefficients is vitally important because the cost of the fermentation medium, particularly the carbon source, can be a significant proportion of the overall production cost.

By performing a range of experiments under different operating conditions, varying medium constituents and component concentrations, pH, temperature, etc., optimum growth/ production conditions can be established. It is also important to determine the maximum specific growth rate (mmax) of the production organism.

This is particularly true for primary metabolites, where product formation is related to growth. To optimize the overall productivity of the system, the microorganism must usually be grown at its mmax.

As previously stated, the operating substrate concentration has a major effect on the growth rate of a micro-organism. By performing a series of batch fermentations, each with a different initial concentration of the limiting substrate, the specific growth rate (m) for each experiment can be determined. These data can then be used to estimate both mmax and the saturation constant *(Ks)* by simply taking the reciprocal values in the Monod equation and rearranging equation to give:

$1 = Ks + S$

$\mu \ \mu_{max}S$

or

$1 = Ks + S$

$\mu \ \mu_{max}S \ \mu_{max}S$

and

$1 = Ks \times 1 + S$

$\mu \ \mu_{max}S \ \mu_{max}S$

A plot of 1/m against 1/*S* should produce a straight line with the intercept on the y-axis at $1/\mu_{max}$ and a gradient equal to $K/\ \mu_{max}$. Therefore, an the key kinetic parameters can be readily determined and when the values of *K*, mmax and Y are known, a complete quantitative description can be given of the growth events occurring during a batch culture.

CONTINUOUS GROWTH KINETICS

Continuous culture fermentations have many applications for both laboratory research and industrial-scale processes. Studies can be performed on all aspects of cell growth, physiology and biochemistry. They are useful for ecological studies and as a genetics tool for the examination of mutation rates, mutagenic effects, etc.

Application in industrial fermentations overcomes many limitations of batch processes. Initially, continuous fermentations start as batch cultures, but exponential growth can then be extended indefinitely, in theory, through the continuous addition of fresh fermentation medium. The reactor is continuously stirred and a constant volume is maintained by incorporating an overflow weir or other levelling device.

Fresh medium is continuously added and displaces an equal volume of spent fermentation broth and cells at the same rate as fresh medium is introduced. Steadystate conditions prevail, where the rate of microbial cell growth equals the rate at which the cells are displaced from the vessel.

As with batch fermentations, the specific rate at, which the microorganism grows in continuous culture is controlled by the availability of the rate-limiting nutrient. Therefore, the rate of addition of fresh medium controls the rate at which the microorganisms grow.

However, the actual rate of growth depends not only on the volumetric flow rate of the medium into the reactor, but also on the dilution rate *(D)*. This equals the number of reactor volumes passing through the reactor per unit time and is expressed in units of reciprocal time, per hour.

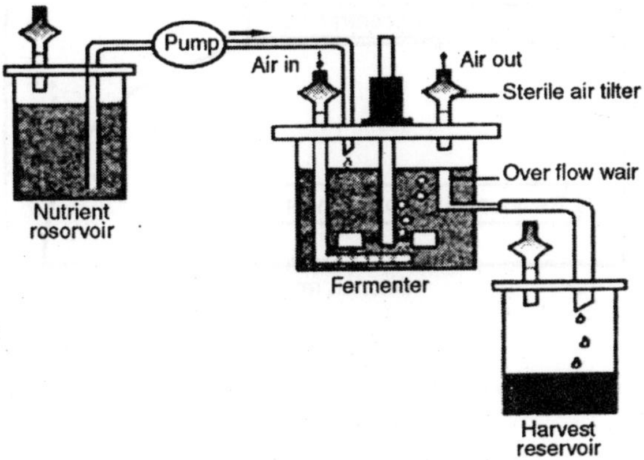

Fig. Continuous Culture Apparatus.

$$D = F\ V$$

where D = dilution rate (per hour), F = flow (L/h) and V= reactor volume (L).

The term D is the reciprocal of the mean residence time or hydraulic retention time, as used in waste-water treatment.Addition of fresh medium into the reactor can be controlled at a fixed value, therefore the rate of addition of the rate-limiting nutrient is constant. Within certain limits, the growth rate and the rate of loss of cells from the fermenter will be determined by the rate of medium input.

Therefore, under steady state conditions the net biomass balance can be described as:

dx = rate of growth-rate of loss from reactor
dt in reactor (wash-out)

or

$$\frac{dx}{dt} = \mu x - Dx$$

Under steady-state conditions the rate of growth = rate of loss,

hence dx/dt = 0 and therefore

μx = Dx

and m = D

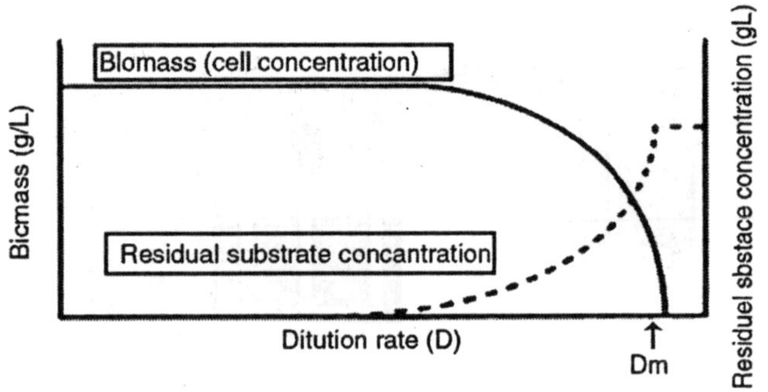

Fig. Growth of a Microorganism in Continuous Chemostat culture.

Consequently, at fixed flow rates and dilution rates under constant physical and chemical operating conditions, i.e. under steady-state conditions, the specific growth rate of the microorganism is dependent on the operating dilution rate, up to a maximum value equal to μ_{max}. If the dilution rate is increased above μ_{max} complete wash-out of the cells occurs, as the cells have insufficient time to 'double' before being washed out of the reactor via the overflow. The point at which this is just avoided is referred to as the critical dilution rate (Dcri). For any given dilution rate, under steady-state conditions, the residual substrate concentration in the reactor can be predicted by substituting D for/*l* in the Monod equation:

D = μ_{max} Sr

Ks + Sr

where Sr is the steady-state residual substrate concentration in the reactor at a fixed dilution rate. Rearrangment gives

D(*Ks* + Sr) = μ_{max} Sr or dividing by Sr then gives

D Ks + D Sr = μ_{max} Sr

D Ks + D = μ_{max} Sr

Sr

Hence,

$$Sr = \frac{DKs}{\mu_{max} - D}$$

Consequently, the residual substrate concentration in the reactor is controlled by the dilution rate. Any alteration to this dilution rate results in a change in the growth rate of the cells that will be dependent on substrate availability at the new dilution rate.

Thus, growth is controlled by the availability of a rate-limiting nutrient. This system, where the concentration of the ratelimiting nutrient entering the system is fixed, is often described as a chemostat, as opposed to operation as a turbidostat, where nutrients in the medium are not limiting. In this case turbidity or absorbance of the culture is monitored and maintained at a constant value by regulating the dilution rate, i.e. cell concentration is held constant.

At low dilution rates with fixed substrate concentrations, the residual substrate concentration will be low. However, as D approaches μ_{max} the residual substrate concentration increases along with the growth rate of the microorganism. Beyond Dcrit, input substrate concentration will equal output concentration, as all the cells have been lost from the system. Thus, this continuous reactor can be described as a selfregulating nutrient-limited chemostat. The concentration of biomass or microbial metabolite in a continuous fermenter under steady-state conditions can be related to the yield coefficient, as described in the batch fermentation section. Inserting the equation for residual substrate into the biomass or metabolic product yield coefficient equation gives, in this case for steadystate biomass *(x)*,

$$\bar{x} = Yx/s\ SR - \frac{DKs}{\mu_{max} - D}$$

where SR is the substrate concentration of inflowing medium or

$$\bar{x} = Yx/S(SR\text{-}Sr)$$

Therefore, the biomass concentration under steady state conditions is controlled by the substrate feed concentration and the operating dilution rate. Under non-inhibitory conditions, where there is no substrate or product inhibition, the higher the feed concentration, the greater the biomass concentration and residual substrate concentration remains constant. However, the higher the dilution rate, the faster the cells grow, which results in a simultaneous increase in the residual substrate concentration and a consequent reduction in the steady-state biomass concentration. As D approaches μ_{max}, the biomass concentration becomes even lower, yet the cells grow faster and there is a concurrent increase in the residual substrate concentration.

MICROBIAL GROWTH AND SUBSTRATE UTILIZATION KINETICS

Contamination of the environment with hazardous and toxic chemicals is one of the major problems facing the industrialized nations today. The petroleum industry is responsible for the generation of high amounts of petroleum hydrocarbons and their derivatives as well as for the pollution of air, soils, rivers, seas and underground water. These compounds undergo modifications by either physico-chemical or biological processes.

Diverse metabolic capabilities of microorganisms have been exploited by man in diverse ways in the biodegradation of waste materials. Microbial activities allowed the mineralization of some petroleum components into carbon dioxide and water, and microbial transformation is considered a major route for complete degradation of petroleum components. The potentiality of microbes as agents of degradation of several compounds thus indicates biological treatment as the major promising alternative to attenuate environmental impact caused by pollutants. Many scientific approaches have been used in the *in situ* and *ex situ* biodegradation of organic pollutants.

However, the extent of biodegradation is critically dependent on salinity, temperature, pH, heavy metals surfactants, nutrients and presence of readily assimilable carbon sources. Many methods such as oxidation, precipitation, ion exchange, solvent extraction, enzyme treatment and adsorption have been used for removing both organic and inorganic materials from aqueous and non-aqueous solution.

A variety of microbial growth and biodegradation kinetic models have been developed, proposed and used by many researchers. Such models allow prediction of chemicals that remain at a certain time, calculation of the time required to reduce chemical to certain concentration, estimation of how long it will take before a certain chemical concentration will be attained at a certain point (e.g. a case of aquifer, soil or surface water) and design of bioremediation schemes *in situ* or *ex situ* to remove chemical contaminant to a designed concentration.

On the other hand, it can be used to predict the amount of biomass production achievable at a given time. This review gives an overview of the kinetic models as applied in the prediction of microbial growth and degradation of organic substances. Substrate inhibition and interactions during biodegradation of pollutant mixtures are also discussed.

MICROBIAL GROWTH KINETICS

The relation between the specific growth rate (m) of a population of microorganisms and the substrate concentration (S) is a valuable tool in biotechnology. This relationship is represented by a set of empirically derived rate laws referred to as theoretical models. These models are nothing but mathematical expressions generated to describe the behaviour of a given system. The classical models, which have been applied to microbial population growth, include the Verhulst and Gompertz function. The Gompertz function was originally formulated for actuarial science for fitting human mortality data but it has also been applied deterministically to organ growth.

The Gompertz function is based on an exponential relationship between specific growth rate and population density.

Equation represents one of its parameterization.

$$N_{(t)} = C\exp\{\exp[-B(t-M)]\}$$

where t = time, N(t)= population density at time t, C = upper asymptotic value, that is; the maximum population density, M = time at which the absolute growth rate is maximal, and B = relative growth rate at M. time.

Gibson et al. modified the Gompertz function to a function which could be applied to the description of cell density versus time in bacterial growth curves in terms of exponential growth rates and lag phase duration:

$$\text{Log}\, N_{(t)} = A + D\exp\{-\exp[-B(t-M)]\}$$

Where N(t) = population density at time t, A = value of the lower asymptote (Log N(–¥)), D = difference in value of the upper and lower asymptote [Log N(¥) – log N(-¥)], M = time at which the exponential growth rate is maximal. The idea of microbial growth kinetics has been dominated by an empirical model originally proposed by Monod. The Monod model introduced the concept of a growth limiting substrate.

$$\mu = \mu_{max} \frac{S}{K_s + S}$$

Where m = specific growth rate, mmax = maximum specific growth rate, S = substrate concentration, Ks = substrate saturation constant (i.e. substrate concentration at half mmax). In Monod's model, the growth rate is related to the concentration of a single growth-limiting substrate through the parameters mmax and Ks. In addition to this, Monod also related the yield coefficient (Yx/s) to the specific rate of biomass growth (m) and the specific rate of substrate utilization (q).

$$Y_{x/s} = \frac{dx}{ds}\mu = \frac{Y_{x/s}}{X} \cdot \frac{ds}{dt} \cong Y_{x/s}q$$

DERIVATIVES OF THE MONOD KINETIC MODEL

In 1912, Penfold and Norris proposed the first kinetic principle for microbial growth. They stated that the relationship between m and S is best described by a "saturation" type of curve where at high concentration of substrate, the organism grows at a maximum rate (mmax) independent of the substrate concentration.

Monod's model satisfies this requirement, but it has been criticized particularly because of derivations of m at low substrate concentration.

Owing to the limitations of the Monod's model, a number of structured and unstructured kinetic expressions were put forward to describe the hyperbolic curve characteristic of microbial growth. However, the development of structured models had suffered serious setback due to the complexity of cell growth. Thus, most proposed growth models are unstructured.

Three approaches were used to develop the equations for growth kinetics of cells in suspension:

- Describing the influence of physicochemical factors on Monod growth parameters.
- Inclusion of additional constants into the original Monod model to correct for substrate or product inhibition, substrate diffusion, maintenance or effects of cell density on mmax.
- Proposing different kinetic theories, which result in both empirical and mechanistic models.

Like the Monod kinetics, Gompertz function has also been modified to generate models that describe the effect of intrinsic factors such as temperature and oxygen availability on microbial growth parameters.

Studies have been carried out on the combined effects of several controlling factors on bacterial growth for example papers by Sutherland et al, McMeekin et al, Wijtzes et al. and Adams et al. These models are mostly used to predict the change in quality of a food over time and can therefore be applied to estimate the shelf-life of foods.

KINETICS OF BIODEGRADATION

The basic hypothesis of biodegradation kinetics is that substrates are consumed via catalyzed reactions carried out only by the organisms with tne requisite enzymes.

Therefore, rates of substrate degradation are generally

proportional to the catalyst concentration (concentration of organisms able to degrade the substrate) and dependent on substrate concentration characteristic of saturation kinetics (e.g. Michaelis-Menten and Monod kinetics).

Saturation kinetics suggests that at low substrate concentrations (relative to the half-saturation constant), rates are approximately proportional to substrate concentration (first order in substrate concentration), while at high substrate concentrations, rates are independent of substrate concentration (zeroorder in substrate concentration).

In the case of substrates that contribute to the growth of the organisms, rates of substrate degradation are linked to rates of growth (i.e. the concentration of the biomass increases with substrate depletion). The mathematical analysis of such growth-linked systems is more complex than those situations where growth can be ignored.

There are a number of situations where it may not be possible to quantify the concentration of substrate-degrading organisms in a heterogeneous microbial community. However, the rate of substrate depletion can be measured. There are also situations in which the organism concentration remains essentially constant even as the substrate is degraded (i.e. no growth situation).

Given these various features of biodegradation kinetics, different models including first-order, zero-order, logistic, Monod (with and without growth) and logarithmic models can be used to describe biodegradation.

Biodegradation kinetics is used to predict concentrations of chemical substances remaining at a given time during *ex situ* and *in situ* bioremediation processes. In most cases, information is based on loss of parent molecule targeted in the process. The key interest is frequently the decrease in toxicity concentration.

Nevertheless, toxicity measurements require bioassays, which are always very difficult and tedious. Therefore, efficacy of biodegradation is based on chemical measurements, e.g. disappearance of parent molecule, appearance of mineralization products or disappearance of other compounds used stoichiometrically during biodegradation of a compound, for instance, electron acceptors. There are several scenarios by which a compound can be transformed biologically.

This includes when the compounds serve as:

- Carbon and energy source
- Electron acceptor
- Source of other cell components.

Other scenarios are the transformation of a compound by non-growing cells (the compound does not support growth) and the transformation of a compound by cometabolism, that is; transformation of a compound by cells growing on other substrate.

The simplest case is where the compound serves as source of carbon and energy for the growth of a single bacterial species. The compound is assumed to be water-soluble, non-toxic and other substrates or growth factors are limiting. In the case of single-substrate limited process, the Monod equation is often used to describe microbial growth and biodegradation processes.

$$\mu = \frac{\mu_{max} S}{K_s + S}$$

$$q = \frac{q_{max} S}{K_s + S}$$

where μ = specific growth rate (1/X.dX/dt), q = specific substrate utilization/removal rate (1/X.dS/dt), and $\mu = Yq$, with Y = true growth yield [mass of biomass (X) synthesized per unit of substrate (S) utilized or removed], S = aqueous phase concentration of the compound, Ks = affinity constant or half saturation constant for the compound (meaning the concentration of compound when m or q is maximum).

The hyperbolic equation proposed by Monod was modified by Lawrence and McCarty to describe the effects of substrate concentration (S) on the rate at which a given microbial concentration (X) removes the target substrate (-dS/dt). Alternatively, Monod equation can be written in terms of microbial growth by incorporating the net yield coefficient (Y).

$$\frac{dS}{dt} = -\frac{q_{max} SX}{k_s + S}$$

$$\frac{dX}{dt} = -Y\frac{dS}{dt} = \frac{Yq_{max} SX}{K_s + S}$$

The Monod equation has frequently been simplified to an equation, which is either zero or first order in substrate concentration and the kinetics, has been widely used to describe biodegradation of organic contaminations in aquifer systems. The versatility of Monod's equation is attributed to its ability to describe biodegradation rates that follow zero-to first-order kinetics with respect to the concentration of the target substrate. Moreso, Monod's model describes the dependence of biodegradation rate on the concentration of biomass.

SUBSTRATE INHIBITION OF BIODEGRADATION

When a substrate inhibits its own biodegradation, the original Monod model becomes unsatisfactory. In this case, Monod derivatives that provided corrections for substrate inhibition (by incorporating the inhibition constant Ki) can be used to describe the growth-linked biodegradation kinetics. Among the substrate inhibition models, the Andrew's equation is most widely used. It is also a good representation of experimental data sets examined in the study of Goudar et al.

$$\mu = \mu_{max} \frac{S}{K_s + S + \frac{S^2}{K_i}} \cdot \quad q = q_{max} \frac{S}{K_s + S + \frac{S^2}{K_i}}$$

A generalized Monod type model originally proposed by Han and Levenspiel has been used to account for substrate stimulation at low concentration and substrate inhibition at high concentration.

$$q = \frac{q_{max}\left(1 - \frac{S}{S_m}\right)^n}{S + K_s - \left(1 - \frac{S}{S_m}\right)^m}$$

Where μ = specific substrate consumption rate of cells, q_{max} = maximum consumption rate constant, S = substrate concentration, K_s = the Monod constant, S_m = critical inhibitor concentration above which reaction stops, n and m are constants. Information available on the substrate inhibition of biodegradation are mostly those that described microbial degradation of phenol. Thus in this review, substrate inhibition of biodegradation is discussed with particular reference to phenol. The inhibitory nature of phenol at high concentrations is well known, and the kinetics of pure and mixed culture microbial growth on phenol have been described by a variety of substrate inhibition models. Most of these models are empirical. However, they are able to provide satisfactory description of phenol biodegradation data, thus providing a convenient means of modelling phenol biodegradation.

Rozich et al. examined 113 microbial curves and reported that among 5 different models, Andrew's provided the best description of observed data. However, the superiority of the Andrew's equation is not a consistent feature in literatures. Pawlowsky and Howell observed statistically insignificant difference between 5 inhibition models. Yang and Humphrey made similar observation with Andrew's equation and 2 other models in describing phenol degradation by *Pseudomonas putida* and *Trichosporon cutaneum*. A two-parameter derivative of Andrew's equation was reported to be a better representation of experimental data obtained from mixed culture biodegradation of phenol.

When different substrate inhibition models are used to describe experimental data, it becomes difficult to compare kinetic parameters across different studies. This complicates the application of laboratory kinetic information in the design of biological treatment systems for inhibitory waste. In the study of Goudar et al, a theoretical basis for selection of an appropriate substrate inhibition model (to solve this problem) was discussed. In this regard, the generalized substrate inhibition model (GSIM) of Tan et al, which describes substrate inhibition of microbial growth using a statistical thermo-dynamics, was used.

ESTIMATION OF MODEL KINETIC PARAMETERS

Kinetic equations, which describe the activity of an enzyme or a microorganism on a particular substrate, are crucial in understanding many phenomena in biotechnological processes. Quantitative experimental data is required for the design and optimization of biological transformation processes. A variety of mathematical models have been proposed to describe the dynamics of metabolism of compounds exposed to pure cultures of microorgainsms or microbial populations of natural environment. The Monod equation has been widely used to describe growth-linked substrate utilization. Characterization of the enzyme or microbe-substrate interactions involves estimation of several parameters in the kinetic models from experimental data. In order to describe the true behaviour of the system, it is important to obtain accurate estimates of the kinetic parameters in these models.

Both derivative and integrated forms of equations derived for enzyme catalyzed reactions have been used to estimate kinetic parameters of microbiological processes. Estimates of kinetic parameters Vmax and Km have been calculated by fitting data to either integrated or derivative forms of Michaelis-Menten and Monod equation. Different approaches have been proposed for estimating the kinetic parameters, but progress curve analysis is the most popular because substrate depletion or product formation data from a single experiment are enough for parameter estimation. In this approach, substrate depletion or product formation-time course is used in the integrated form of the kinetic model for parameter estimation. Some of these differential and integral equations can be found in the papers of Gouder and Delvin, Schmidt et al. and Simkins and Alexander.

It is important to note that most kinetic models and their integrated forms are nonlinear. This makes parameter estimation relatively difficult. However, some of these models can be linearized. Various linearized forms of the integrated expressions have been used for parameter estimation.

However, the use of linearized expression is limited because it transforms the error associated with the dependent variable making it not to be normally distributed, thus inaccurate parameter estimates.

Therefore, nonlinear least-squares regression is often used to estimate kinetic parameters from nonlinear expressions. However, the application of nonlinear least-squares regression to the integrated forms of the kinetic expressions is complicated.

This problem and solutions were discussed by Goudar and Delvin. The parameter estimates obtained from the linearized kinetic expressions can be used as initial estimates in the iterative nonlinear least-squares regression using the Levenberg-Marquardt method.

The kinetic parameters of the Andrew's equation (mmax, qmax, Ks or Ki) can be estimated with the application of reduced form of the generalized substrate inhibition model (GLIM), reduced to the form of Andrews equation. The linearized expression of this model was used to obtain initial parameter estimates for use in nonlinear regression.

SUBSTRATE INTERACTIONS

Wastewaters from industrial and municipal sources are characterized by presence of mixtures of chemicals. Pollutant mixtures may contain only organic chemicals or may also include inorganic substances such as heavy metals. Co-contamination of natural environments with mixtures of pollutants is an important problem. In biodegradation or bioremediation investigations and projects, it is important to understand and be able to model the fate of specific chemicals.

Development of treatment strategies for soil or water contamination requires consideration of interactions among substrates to control the concentration of individual pollutants to meet regulatory standards. Single substrate kinetic parameters alone cannot describe the phenomena observed with degradation of mixtures. It is important therefore to predict the biodegradation kinetics of pollutant mixtures in a given system.

The removal of one component may be inhibited by other components in the mixture and different conditions may be required to degrade different compounds within the mixture. Biodegradation patterns of a compound as component of pollutant mixture and as a single component have been shown to be different. Strong interactions among components of a pollutant mixture have been reported.

In the case of homologous mixture (mixture of substrates serving the same purpose) of carbon and energy substrates, the effect of other compounds in a mixture can be positive or negative due to competitive, and the formation of toxic intermediates by non-specific enzymes.

The utilization pattern can change with different mixture compositions, depending on the chemical nature and concentration of the substrate, oxygen concentration and microbial growth rates.

Arvin et al. observed both substrate inhibition and stimulation interactions during the aerobic degradation of mixtures of benzene, toluene and *o*-xylene. When toluene or *o*-xylene was degraded in the presence of benzene, the degradative ability of toluene and *o*-xylene by the microorganisms was stimulated. When *p*-xylene and toluene were both present, an inhibition effect on benzene degradation was observed.

Similar inhibition and stimulation of biodegradation have been observed with mixtures of benzene, toluene and *p*xylene. In addition to biodegradation stimulation due to increased growth at low substrate concentrations, stimulation of one compound by another in a mixture can be by induction of catabolic enzymes required for degradation of the second pollutant.

This mechanism produces simultaneous degradation of pollutants in mixtures and has been reported for pentachlorophenol and chlorinated aromatics, toluene and *p*-xylene, and toluene. Moreso, one component of a mixture can be degraded in the presence of another by co-metabolism.

In the literature, most studies on the kinetics of biodegradation were on single substrate utilization. However, models of mixed homologous substrate utilization and microbial growth have been proposed. Most of these models have been tested with only two substrates. However, in recent times, models have been proposed and tested for larger mixtures.

Typical examples include the growth of *Escherichia coli* on six sugars, the growth of a mixed culture on five BTEX compounds, and the biodegradation of three polycyclic aromatic hydrocarbons. Like with the homologous substrates, some efforts have also been made to develop kinetic models that described multiple-nutrient-controlled growth with heterologous substrate (substrate that serves different purposes, e.g. carbon and nitrogen mixtures). The heterologous substrate concept assumes that the growth rate can be affected simultaneously by more than one substrate. A "Double Monod" model originally proposed by McGee et al. was used to describe this phenomenon.

$$\mu = \mu_{max} \frac{S_1}{K_1 S_1} \frac{S_2}{K_s + S_2}$$

Where 1 and 2 represent the substrates. However, this multiplicative model has narrow range of utility. Mankad and Bungay had expressed growth rates under dual substrate limitation in terms of weighted average of rates under individual nutrient limitations.

$$\frac{\mu}{\mu_{max}} = (W1)\frac{S_1}{K_1 + S_1} + (W2)\frac{S_2}{k_2 + S_2}$$

Where W (*i*) is the weight assigned to nutrient *i*. substituting the functional dependence for the weight functions W (*i*) into equation yields the Mankad and Bungay's expression for growth rate.

$$W1 = \frac{\frac{K_1}{S_1}}{\frac{K_1}{S_1} + \frac{K_2}{S_2}}; W2 = \frac{\frac{K_2}{S_2}}{\frac{K_1}{S_1} + \frac{K_2}{S_2}}$$

$$\frac{\mu}{\mu_{max}}=\frac{\frac{K_1}{S_1}}{\frac{K_1}{S_1}+\frac{K_2}{S_2}}\left(\frac{S_1}{K_1+S_1}\right)+\frac{\frac{K_2}{S_2}}{\frac{K_1}{S_1}+\frac{K_2}{S_2}}\left(\frac{S_2}{K_2+S_2}\right)$$

For the homologous substrate, sum kinetic model incorporating purely competitive substrate kinetics was proposed by Yoon et al.

$$\mu(S_1,S_2)=\mu_{max}\frac{\mu_{max,1}S_1}{K_{S,1}+S_1+\left(\frac{K_{S,1}}{K_{S,2}}\right)S_2}+\frac{\mu_{ma,2x}S_2}{K_{s2}+S_2+\left(\frac{K_{S,2}}{K_{s,1}}\right)S_1}$$

Equation indicates that each substrate exhibits a competitive inhibition effect on the utilization of the other substrate. The competitive substrate kinetics can be used to describe simultaneous and sequential substrate consumption for mixtures of substrate.

Another form of dual-substrate interaction with an enzyme is noncompetitive inhibition, characterized by the formation of a non-reactive complex when both substrates are simultaneously bound to the enzyme. The cell growth model based on this type of interaction is expressed mathematically.

$$\mu=\frac{\mu_{max1}S_1}{(K_{S1}+S_1)\left(1+\frac{S_2}{ks2}\right)}+\frac{\mu_{max2}S_2}{(K_s2+S_2)\left(1+\frac{S_1}{K_{S1}}\right)}$$

Uncompetitive enzyme inhibition model has also been used to describe dual substrate interaction. It differs from non-competitive inhibition in that one of the compounds (the inhibitor) can bind only to the enzyme substrate complex and not the free enzyme. A cell growth model based on uncompetitive substrate interaction is:

$$\mu=\frac{\mu_{max1}S_1}{K_{s1}+S_1\left(1+\frac{S_2}{K_{s2}}\right)}+\frac{\mu_{max2}S_2}{K_{s2}+S_2\left(1+\frac{S_1}{K_{S1}}\right)}$$

In the sum kinetic models, kinetic parameters determined

in the single substrate experiments are used for curve fitting. These models were evaluated by Reardon et al. for biodegradation of benzene, toluene and phenol mixtures, and found that the interactions between these substrates could not be described by sum kinetics models using only parameters determined in a single substrate experiment. An alternative model was formulated by adding an unspecified type of interaction into the sum kinetics model to produce the sum kinetics with interaction parameter (SKIP) model first proposed by Yoon et al.

$$\mu = \frac{\mu_{max1} S_1}{K_{S1} + S_1 + 1_{2,1} S_1} + \frac{\mu_{max2} S_2}{k_{s2} + S_2 1_{1,2} S_1}$$

The interaction parameter Ii, j indicates the degree to which substrate i affects the biodegradation of substrate j. The larger the value, the stronger the inhibition. The SKIP model form for a three-compound mixture is:

$$\mu = \left[\frac{\mu_{max1} S_1}{K_{s3} + S_1 + 1_{2,1} S_2 + I_{3,1} S_3}\right] + \left[\frac{\mu_{max2} S_2}{K_{s2} + S_2 I_{1,2} S_1 + I_{3,2} S_3}\right] + \left[\frac{\mu_{max3} S_3}{K_{S3} + S_3 I_{1,3} S_1 + I_{2,3} S_2}\right]$$

where the subscripts 1, 2, and 3 denote parameters for three different substrates. The extended SKIP model for N substrates is expressed as:

$$\mu = (S_1 S_2 \ldots, S_N) = \sum_{i=1}^{N} \frac{\mu_{max,i} S_i}{K_{si} + S_i + \sum_{j=1,\, j=1}^{N} S_j I_{j.i}}.$$

The effect of one substrate on the degradation of another is given by the SjIj,i terms. The values of the interaction coefficients, Ij,i, represent the degree of inhibition exerted by substrate j on substrate i. In a dualsubstrate system, sequential substrate utilization is represented by a large value of I1, 2 and a small value of I 2, 1. The SKIP model satisfactorily described simultaneous and sequential degradation patterns in two different biological systems.

PESTICIDE BIODEGRADATION KINETICS

Increased agricultural practice and pesticide application had resulted in the contamination of natural environments with different kinds of pesticides. Wolt et al. described the design and interpretation of biodegradation studies conducted globally for the purpose of regulatory decision making with respect to pesticide use.

Emphasis was placed on the various approaches utilized for addressing degradation studies in soil and the variability in pesticide soil fate parameters. Understanding pesticide risks requires characterizing pesticide exposure within the environment in a manner that can be broadly generalized across widely varied and soil degradation are especially important for understanding the potential environmental exposure of pesticides. The data obtained from degradation studies are inherently variable and when limited in extent, lend uncertainty to exposure characterization and risk assessment. Pesticide decline in soils reflects dynamically coupled processes of sorption and degradation that add complexity to the treatment of soil biodegradation data from a kinetic perspective.

Additional complexity arises from study design limitations that may not fully account for the decline in microbial activity of test systems or that may be inadequate for considerations of all potential dissipation routes for a given pesticide. Accordingly, kinetic treatment of data must accommodate a variety of differing approaches starting with very simple assumptions as to reaction dynamics and extending to more involved treatments if warranted by the available experimental data.

Selection of the appropriate kinetic model to describe pesticide degradation should rely on statistical evaluation of the data fit to ensure that the models used are not over parameterized. Recognizing the effects of experimental conditions and methods for kinetic treatment of degradation data is critical for making appropriate comparisions among pesticide biodegradation data sets. Statistical evaluation of measures of central tendency for multisoil kinetic studies shows that geometric means better represent the distribution in soil half-lives than do the arithmetic or harmonic means.

METAL INHIBITION OF BIODEGRADATION AND PREDICTION OF METAL SPECIATION

In sites co-contaminated with metals and organic compounds, metal toxicity inhibits the activity of organicdegrading microorganisms, impacting both their physiology and ecology, thus reducing the rate of biodegradation of the organic compounds.

Metal inhibition of a broad range of microbial processes including methane metabolism, growth, nitrogen and sulphur conversions, dehalogenation and reductive processes in general is well documented. Thorough reviews of the impacts of metals on many of these processes are available.

The toxicity of metals to microorganisms is dependent on its bioavailabity. Quantification of bioavailable metal concentration is an important step in the process of standardizing experiments to determine the impact of metals on organic pollutant biodegradation. Concentrations of bioavailable metals (metal speciation) can be estimated from solution phase using ion-selective electrodes and atomic absorption spectroscopy. Biological systems involving immunoassay or bioreporters have been used for mercury.

However, the use of immunoassay and bioreporters is limited because of variation in measurements depending on the metal resistance of the bioreporter system used. The application and limitations of immunoassay and bioreporters for metal detection have been reviewed by Neilson and Maier. As the alternative, bioavailable metal concentrations as a function of pH and ionic strength can be predicted using geochemical modeling software's. A number of computational models have been developed to predict the impact of metal on organic biodegradation.

These models accounted for metal inhibition by incorporating metal inhibition constant (Ki) to conventional growth or degradation model. For example, Amor et al. used a form of the Andrew's equation (originally used to describe substrate inhibition of microbial growth or substrate degradation) to model the effect of cadmium, zinc and nickel on rates of alkyl benzene biodegradation.

BIOREACTOR

A bioreactor may refer to any device or system that supports a biologically active environment. In one case, a bioreactor is a vessel in which is carried out a chemical process which involves organisms or biochemically active substances derived from such organisms. This process can either be aerobic or anaerobic. These bioreactors are commonly cylindrical, ranging in size from liters to cubic meters, and are often made of stainless steel.

BIOREACTORS IN BIOTECHNOLOGY

Bioreactors are the places where proteins for biotechnology are made during upstream processing. Today, bioreactors can be vessels made of plastic, glass or steel in which cells are cultured or, alternatively, whole plants or animals can be genetically engineered to produce a particular protein (transgenics). In this discussion we will focus on the types and characteristics of the glass, steel or plastic bioreactors used to house suspension or fixed cells for the purpose of making proteins during upstream processing.

Bioreactor vessels (also referred to as fermenters) can be almost any vessel capable of holding cells. In this course we will be using 50ml and/or 500ml glass Bellco Spinner Bottles with a plastic cap for holding a shaft with an impeller for mixing and a glass 5000ml (5liter) Bioflo 3000 bioreactor with a stainless steel headplate for holding a shaft with an impeller for mixing, various sensors, a sparging tube and a sampling port for culturing suspension cells for upstream processing of proteins. In addition, during the scale-up of bacterial and yeast cell cultures for upstream processing, we will also be using shake flasks and/or test tubes which also could be termed "bioreactors". Other simple bioreactors include t-flasks and roller bottles.

Once assembled, the bioreactor is sterilized either before or after the addition of media. The sterilization is either in place (SIP) or the bioreactor is placed in an autoclave and sterilized there. Following sterilization of the first bioreactor used in the scale-up process, one milliliter of cells to be cultured is added to the media in this bioreactor. Typically, the one milliliter of cells comes from a master cell bank laid down during process development and stored in liquid nitrogen or in a-86 degree Centigrade freezer. This one milliliter of cells from the master cell bank ordinarily contains about 1 million live cells. After these cells have grown and multiplied a number of generations so that the concentration of cells reaches 1 million cells/ml again, this batch of cells is added to a larger bioreactor vessel containing a larger volume of media. Scale-up continues until the cells reach the final bioreactor where the final stage of upstream processing occurs.

The media or the cells are then harvested for the downstream processing of proteins produced during culture in the final bioreactor. The final bioreactor used in industry for upstream processing of proteins is far larger than the final bioreactors used in this class. Typically, for the production of human therapeutic proteins, a stainless steel bioreactor constructed of the best quality (316) virgin steel with a working volume of 2000 to 5000 liters is used. However, stainless steel bioreactors with working volumes of up to 100,000 liters are also in use today. The most common shape for these stainless steel bioreactors is a vessel that is about as wide as it is tall containing a Rushton impeller which looks a bit like a boat's propeller. This is a stirred tank bioreactor. An alternative to this shape is the airlift bioreactor which is quite a lot longer than it is wide (a kind of bullet shape) that uses air sparged at its bottom as a mixing device.

There are many more alternatives to these typical final bioreactors. At one company, the final bioreactor for the manufacture of the protein, tissue plasminogen activator (tPA), is carried out in multiple 10 liter glass Bellco Spinner Bottles, with previous scale-up steps carried out in 1000ml and 100ml Bellco Spinner Bottles. (Bellco Spinner Bottles are available from 25ml to 36,000ml.)

For industrial upstream processing of proteins various types of plastic bioreactors are also in use including fixed cells in a fluidized-bed bioreactor. In this bioreactor the cells are fixed in a compartment and media is constantly pumped to the cells in the fixed compartment and constantly withdrawn and pooled for downstream processing.

Another plastic bioreactor is the hollow fibre bioreactor in which cells grow inside hollow fibers or tubes and media is pumped either through the hollow fibers or over the surface of the hollow fibers and constantly withdrawn and pooled for downstream processing. Another sort of bioreactor is the perfusion bioreactor. In industry, where quality control in the form of cGMP is a necessity, many parameters of bioreactors must be controlled and recorded. It is typical to use biosensors connected to computers to control and record these parameters. This is referred to as process control.

This allows for the monitoring of various parameters and provides negative feedback to insure that the bioreactor will operate within pre-defined limits. This is not un-like our bodies which have sensors and negative feedback mechanisms to keep us operating in homeostasis (or within pre-determined limits). The physical parameters that can be measured on-line during fermentation or cell culture include pH, temperature, dissolved oxygen (DO), dissolved CO_2, impeller speed (or agiation rate), torque, power, foaming, volume of liquid, CO_2 and O_2 in exit gasses, substrate concentration, product concentration, and biomass concentration (or turbidity).

Some of these parameters are still measured by fermentation or upstream processing operators or technicians, by "hand". For instance, it is common for the operator or technician to remove samples at periodic intervals and check the biomass concentration or turbidity by taking an OD reading on the spectrophotometer and also by staining the cells in the sample with vital dyes to determine the live cell number/

ml of medium. Quality Control technicians often measure the concentration of the protein of interest in such samples. Many sensors are still under development inlcuding those that would measure the concentration of various enzymes, substrates, ions, ATP, DNA and RNA and others.

pH Control Loop:

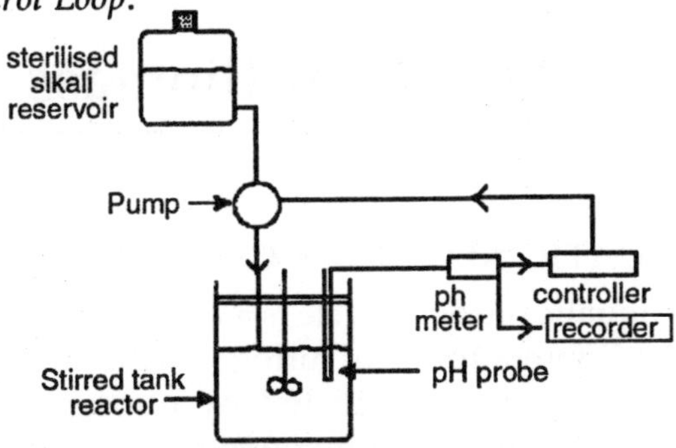

Fig. Dissolved Oxygen Control Loop

In our laboatory, the process controlled New Brunswick Bioflo 3000 bioreactor is connected to a computer loaded with New Brunswick AFS Software. In our yeast culture protocol for the making of human serum albumin (HSA) control of pH, DO, and the addition of media will take place automatically. A schematic representation of a pH control loop and a dissolved oxygen control loop taken from one of your texts is shown above:

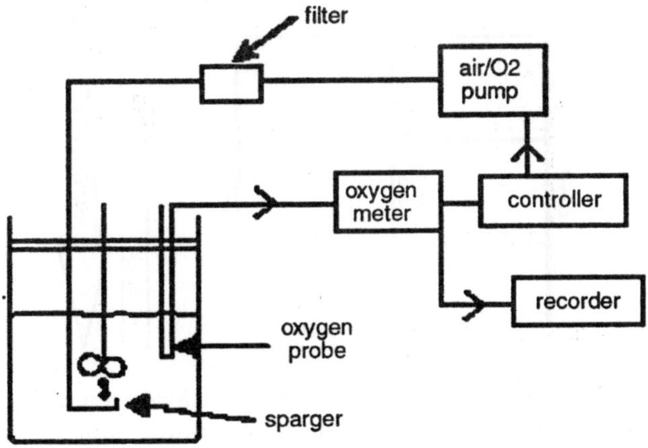

Chapter 5

Microbial Biomass

THE AQUATIC NITROGEN CYCLE

The aquatic nitrogen cycle is a complex system consisting of several different forms of nitrogen with different oxidation states that are interconnected by numerous conversion reactions. The different components of the aquatic nitrogen cycle can be separated from each other in space and/or time. The nitrogen cycle plays an important role in aquatic systems since nitrogen is usually the element limiting microbial production in these systems. However, due to its complexity, various aspects of the nitrogen cycle are still poorly understood.

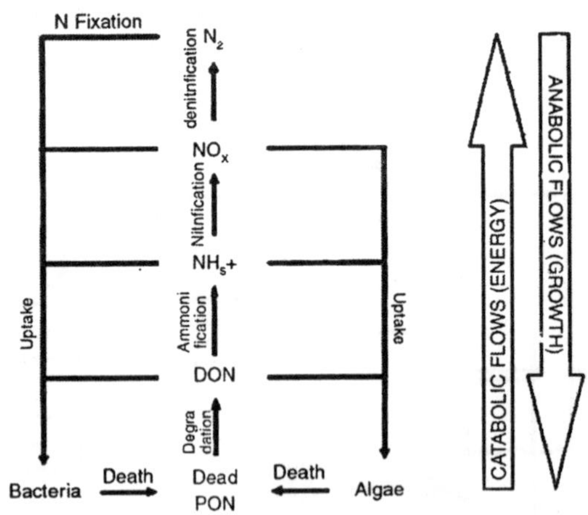

Fig. General Overview of Nitrogen Transformation Processes within the Nitrogen Cycle, Including Nitrogen flows Through Bacteria and algae.

Bioavailable nitrogen is present in the environment as "free" nitrogen (atmospheric N_2) and "fixed" nitrogen. The latter can roughly be divided into particulate nitrogen (functionally defined as all nitrogenous material that can be captured by filtration) and dissolved nitrogen (all nitrogenous material that passes through a filter, typically GF/F).

The dissolved fixed nitrogen pool can be subdivided into dissolved inorganic nitrogen (DIN) and dissolved organic nitrogen (DON). The DIN pool includes ammonium (NH^{4+}), nitrate (NO^{3-}) and nitrite (NO^{2-}) of which the last two can be taken together as NOx-.

The DON pool is a more complex pool of which only a small fraction (< 20%) has been chemically identified. The identifiable fraction mainly consists of urea, dissolved free amino acids (DFAAs) and dissolved combined amino acids (DCAAs, mainly proteins and peptides).

The conversion reactions within the nitrogen cycle can have two "Directions". In catabolic reactions, nitrogenous substrates are used as a source of energy, these reactions are generally mediated by bacteria.

In anabolic (or "assimilatory") reactions, nitrogenous substrates are used as a source of nitrogen for growth (i.e. production of biomass), which requires energy. The uptake and incorporation of various forms of fixed nitrogen by natural aquatic microbial communities.

Microbial Nitrogen Uptake

Natural aquatic microbial communities (mainly bacteria and algae) can use various forms of nitrogen as nitrogen sources for growth.

Although N_2 is abundantly present in the atmosphere, its importance as a nitrogen source for aquatic microbial communities is limited, especially in systems with high availability of fixed nitrogen sources, because the use of nitrogen from N_2 for growth ("Nitrogen fixation") requires a high energy investment and only a limited group of organisms (cyanobacteria and some heterotrophic bacteria) is capable of nitrogen fixation Therefore, the main nitrogen sources for microbial communities in coastal waters and sediments are DIN and DON.

Traditionally, studies on aquatic microbial nitrogen uptake focused on uptake of DIN by algae while bacteria were only considered as remineralizers of organic matter.

However, the present view of microbial nitrogen uptake is considerably more complex since DON (especially amino acids and urea) has been found to be a potentially important nitrogen source as well and because it is now clear that bacteria are also important players in aquatic microbial nitrogen uptake.

In addition to bacteria and algae, aquatic microbial communities can also include Archaea. However, at present their role in total microbial nitrogen uptake is unknown.

The pathways for uptake of dissolved nitrogen by algae and bacteria are schematized. The uptake process consists of two steps that are not necessarily coupled. The first step is the uptake of dissolved nitrogenous compounds from the surrounding water into the cells. Small molecules like,

$$NH_4^+, NH_x^-$$

and urea can easily be taken up into cells while larger compounds like proteins and peptides first need to be broken down into individual amino acids that can either be directly taken up into the cells or be broken down to,

$$NH_4^+$$

which is then taken up into the cells.

The second step is incorporation of the nitrogen in new biomass. For this, intracellular,

$$NH_3^-$$

and urea are first converted to,

$$NH_4^+$$

which is used to synthesize amino acids that form the building blocks for proteinaceous biomass components. Since amino acids can directly be incorporated into biomass after uptake (i.e. little or no conversion is required), these are a highly preferred nitrogen source.

Conversely, the microbial preference for,

$$NH_x^-$$

as a nitrogen source is low because of the relatively high energy investment required for its conversion to,

$$NH^{4+}.$$

Preferences for NH_4^+ and urea are generally in between those for amino acids and,

$$NO^{x-}.$$

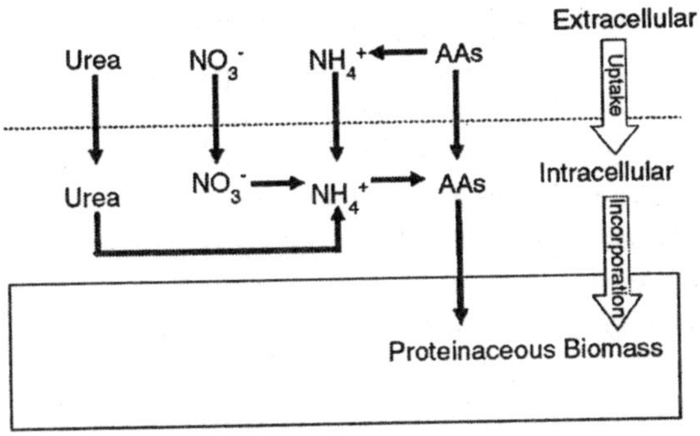

Fate of Microbial Nitrogen

An interesting, yet poorly studied, aspect of the aquatic nitrogen cycle is the fate of microbial nitrogen (i.e., what happens with the nitrogen after it has been incorporated into microbial biomass?) which is coupled with the fate of other elements in microbial biomass such as carbon.

Although nitrogen can be retained in microbial biomass during continued growth (by cell division), microbial biomass (especially that of bacteria) is often relatively constant over time, meaning that production of new biomass is balanced by loss of biomass due to cell death.

Death of microorganisms can be due to various causes such as viral lysis, predation and deteriorating environmental conditions. After cell death, the leftovers (remnants) of microbial biomass are subject to degradation. During degradation, nitrogen is released back into the system as DON and/or DIN that can be used as nitrogen source for production of new microbial biomass or be lost from the system.

Some microbial biomass components are more resistant to degradation ("Refractory") than others and may accumulate in the system where they serve as a long term sink for microbial-derived nitrogen. Another potential fate of microbial nitrogen is transfer to higher trophic levels of the food web when microbial biomass is consumed by larger organisms.

Stable Isotopes in Ecological and Biogeochemical

Elements such as carbon (C) and nitrogen (N) can occur in different forms called isotopes. Different isotopes of a single element have the same number of protons but a different number of neutrons. Some isotopes are radioisotopes, meaning that these isotopes are subject to radioactive decay.

Radioisotopes have been used in ecological and biogeochemical studies (for example for measuring primary production using 14°C). However, work with radioisotopes comes with a potential health risk and can be restricted by law. Moreover, there is no convenient radioisotope of nitrogen which makes that the use of radioisotopes in studies on nitrogen cycling is restricted to work with the short-lived radioisotope ^{13}N

Isotopes that are not subject to radioactive decay are stable isotopes. A single element can have two or more stable isotopes, of which one is by far the most abundant (e.g. 12°C and ^{14}N) while the other forms are rare (for example: 99.6% of all nitrogen in nature is ^{14}N and only 0.4% is ^{15}N). The abundance of these rare isotopes is measured by isotope ratio mass spectrometry (IRMS), a technique that uses a magnetic field to separate gas molecules (CO_2 in case of 13°C analysis and N_2 in case of ^{15}N analysis) of different molecular weights resulting from different stable isotope compositions.

The relative abundance of these molecules with different molecular weights can be used to determine the relative abundance of the rare (or "Heavy") isotope.

This abundance is typically expressed by the delta notation (E) which is the ratio between the rare ("Heavy") isotope and the common (or "Light") isotope relative to that of a standard in permille (%°) units.

Table. Carbon and Nitrogen Isotopes

		Type of Isotope	Natural Abundance
Carbon	^{10}C	Radio (½ life 17 Sec)	
	^{12}C	Stable	98.9%
	^{13}C	Stable	1.1%
	^{13}C	Radio (½ life 5730 yr)	
Nitrogen	^{13}N	Radio (½ life 10 min)	
	^{14}N	Stable	99.6%
	^{15}N	Stable	0.4%
	^{16}N	Radio (½ life 7 Sec)	

Stable isotopes can be used in ecological and biogeochemical studies in two different ways. The first way is to study the natural abundance of the rare, heavy isotope.

Here, the relevant information lies in small differences in relative abundance of the heavy isotope between different compartments of a system. These differences are the results of isotopic fractionation, which involves changes in the natural abundance of the heavy isotope during reactions caused by small differences in mass-dependent reaction rates of heavy and light isotopes. The second way to use stable isotopes is by stable isotope labeling. In this approach, a substrate that is highly enriched in the heavy isotope is deliberately added to a system after which the abundance of the heavy isotope (the "Label") is measured in various components of the system which allows to trace and quantify flows of the label trough these components.

The material presented in this thesis all concerns stable isotope labeling work with the nitrogen isotope ^{15}N and the carbon isotope 13°C.

BIOMARKERS

Biomarkers are compounds that are unique to a specific group of organisms. Ideally, biomarker compounds are part of the organism's biomass and make up a relatively constant fraction of that biomass so that biomarker concentrations can be converted to total biomass of the group it represents. A variety of compounds have been used as biomarkers for various microbial groups.

Some of these compounds are rapidly degraded after death of the organism, which makes them specific for living biomass ("ecological biomarkers"), for example phospholipid-derived fatty acids (PLFAs).

Other components are more resistant to degradation ("biogeochemical biomarkers"), for example D-amino acids from bacterial cell walls, which makes that these components can accumulate in a system, particularly in sediment where the abundance of these biomarkers can provide information on the contribution of material from the corresponding group of organisms to the total organic matter pool in the sediment. A third group of biomarkers are highly refractory molecules ("geological biomarkers"). The concentrations and characteristics of these "molecular fossils" in sediments can be used for paleoenvironmental reconstructions. A valuable application of biomarkers in ecological- and biogeochemical studies is analysis of their stable isotope composition, which requires compound-specific stable isotope analysis.

This can be done at natural abundance level or in combination with stable isotope labeling where the latter provides the unique opportunity to directly link microbial identity (biomarker), biomass (biomarker concentration) and activity (label incorporation). Selection of bacterial biomarkers that have been used in combination with stable isotope analysis. Biomarkers and stable isotope analysis of biomarkers, including some examples of applications in microbial ecology, can be found in Boschker and Middelburg while technical aspects of compound-specific stable isotope analysis.

The Problem

Microbial uptake of nitrogen from various nitrogenous substrates is generally studied by analysis of uptake of ^{15}N from ^{15}N-labeled substrates. This approach has frequently been used in water column studies where ^{15}N enrichment of total particulate material in the water is considered to represent total microbial ^{15}Nuptake.

However, more detailed assessment of ^{15}Nuptake by specific microbial groups (bacteria versus algae) has long been hampered by methodological problems. Until recently, two different approaches have been used to discriminate ^{15}N uptake by bacteria versus algae.

The first approach is size fractionation, which concerns physical separation of relatively small bacteria from relatively large algae by filtration or flow cytometry.

Although this approach may work in clear waters when a size difference between bacteria and algae is present, size fractionation is not suitable for studies in more turbid waters where bacteria and algae are closely associated with particulate matter. The second approach involves the use of inhibitors to eliminate activity of a specific microbial group, for example antibiotics to inhibit bacterial activity. However, the use of inhibitors is seriously limited by uncertainties concerning both the specificity and efficiency of inhibitors.

Next to the problems in water column studies presented above, studies on microbial nitrogen uptake in sediments have been hampered by even more fundamental methodological problems because of the inability to distinguish between ^{15}N taken up by the microbial community and the remaining added ^{15}N-substrate and/or possible other ^{15}N-labeled extracellular compounds bound to the sediment.

Moreover, another limitation of traditional ^{15}N-labeling work is that it does not allow clean assessment of immobilization of dissolved nitrogen into microbial biomass (the relevant process from a nitrogen cycling perspective as well as from a food web point of view) since total ^{15}N "Uptake" also includes uptake of nitrogenous substrates into microbial cells, which is not necessarily directly coupled to incorporation into biomass. The method presented in this thesis is a new approach that has the potential to overcome the problems mentioned above and therefore may help to further clarify nitrogen flows through bacteria and algae in natural microbial communities.

The Solution

The method presented in this thesis involves the use of compound-specific stable isotope analysis of biomarkers. This approach has already been used at NIOO-CEME for some years in 13°C-labeling studies where analysis of 13°C incorporation into various group-specific PLFAs provided valuable information on 13°C flows through different microbial groups, including bacteria and algae. However, since PLFAs do not contain nitrogen, another type of compounds was required for ^{15}N labeling studies.

This thesis investigates the potential of ^{15}N analysis of hydrolysable amino acids (HAAs), including the bacterial biomarker D-alanine (D-Ala).

HAAs are amino acids that can be released from proteinaceous biomass (proteins and protein-like material) by hydrolysis in hot acid.

HAAs make up a large fraction (50- 80%) of total microbial biomass meaning that analysis of ^{15}N incorporation into total HAAs (THAAs) should provide a good indication of ^{15}N incorporation into total microbial biomass. Amino acids can occur as D- and L-stereoisomers.

Most HAAs are L-amino acids that are common biomass constituents of all organisms and therefore have very limited biomarker potential Conversely, hydrolysable D-amino acids are rare and only occur in bacterial biomass, more specifically in peptidoglycan, a structural component of bacterial cell walls.

Therefore, these D-amino acids can be used as bacterial biomarkers. Although different D-amino acids can be present in peptidoglycan, our method focuses on D-Ala because this is the only D-amino acid that is present in all bacteria and because it makes up a relatively stable fraction of total bacterial biomass. Moreover, D-Ala has been shown to be most suitable for stable isotope analysis by GC-c-IRMS. The method can also be used to measure incorporation of the carbon isotope 13°C into D-Ala and other HAAs, which allows assessment of carbon flows through bacteria and algae. This is particularly interesting in direct combination with analysis of nitrogen flows.

Chapter 6

Stem Cells

Tissue engineering is an interdisciplinary field that applies the principles and methods of bioengineering, material science, and life sciences toward the assembly of biologic substitutes that will restore, maintain, and improve tissue functions following damage either by disease or traumatic processes. The general principles of tissue engineering involve combining living cells with a natural/synthetic support or scaffold to build a three-dimensional living construct that is functionally, structurally and mechanically equal to or better than the tissue that is to be replaced.

The development of such a construct requires a careful selection of four key materials:

- Scaffold,
- Growth factors,
- Extra cellular matrix,
- Cells.

Much progress has been made in the field of tissue engineering, but further work toward organ and tissue replacement is necessary. The optimal cell source, scaffold design, and in vitro bioreactors, the use and development of micro fabrication technology to create vascularized tissues and organs are still being investigated.

The search for and use of an appropriate multipotent or pluripotent stem cell in tissue engineering is an emerging concept. Certainly, many areas of stem cell research and their potential clinical applications are associated with controversies; therefore, it is important to address the ethical, legal, and social issues early.

This paper will provide an overview of tissue engineering and stem cells, and describe the current progress with stem cell research in tissue engineering, the potential implications on medical treatment and the economic impact with the passage of Proposition.

Artificial transplantation or transplanted organs is a successful therapy for otherwise incurable end-stage diseases or tissue loss. However, such interventions are challenged by organ shortage, the necessity of lifelong immunosuppression and its potential for serious complications.

Tissue engineering has emerged as a rapidly expanding approach to address these problems and is a major component of regenerative medicine. Tissue engineering is an interdisciplinary field that applies the principles and methods of bioengineering, material science, and life sciences toward the assembly of biologic substitutes that will restore, maintain, and improve tissue functions following damage either by disease or traumatic processes.

The general principles of tissue engineering involve combining living cells with a natural/synthetic support or scaffold that is also biodegradable to build a threedimensional living construct that is functionally, structurally and mechanically equal to or better than the tissue that is to be replaced.

The development of such a construct requires a careful selection of four key components:

- Scaffold,
- Growth factors,
- Extracellular matrix (ECM),
- Cells.

Scaffold materials are three-dimensional tissue structures that guide the organization, growth and differentiation of cells. Scaffolds must be biocompatible and designed to meet both nutritional and biological needs for the specific cell population.

Growth factors are soluble peptides capable of binding cellular receptors and producing either a permissive or preventive cellular response toward differentiation and/or proliferation of tissue. ECM must be capable of providing the optimal conditions for cell adhesion, growth, and differentiation within the construct by creating a system capable of controlling environmental factors such as pH, temperature, oxygen tension, and mechanical forces.

These conditions are determined by the particular cell lines and the properties of the scaffold. Finally, the development of a viable construct involves a suitable supply of cells that are ideally nonimmunogenic, highly proliferative, easy to harvest, and have the ability to differentiate into a variety of cell types with specialized functions.

There are two primary methods to harvest cells and culture. In cases where direct harvest is not feasible, as seen in many patients with extensive end-stage organ failure or cells with limited proliferative capacity in culture, stem cells are envisioned as being an alternative source of cells. Despite the growing interest in embryonic stem cell research, progress has been hampered by ethical and legislative debates.

The political, ethical, and religious opposition toward embryonic stem cell research, which primarily uses discarded nontransferred human embryos for their derivation, have biased research toward adult stem cells and severely restricted federal funding in the United States. In the mist of such controversies, California has become the leading state to support stem cell research by passing Proposition or the California Stem Cell Research and Cures Initiative. It promises to provide California a possible solution to bridge critical research funding gaps by committing $3 billion for stem cell research and to potentially accelerate the understanding of their therapeutic potential.

It does not however support human reproductive cloning. The purpose of this paper is to provide an overview of tissue engineering and stem cells, and describe the current progress with stem cell research in tissue engineering and the potential implications on medical.

FUNDAMENTALS OF TISSUE ENGINEERING

In 1933, the concept of tissue engineering was first introduced when mouse tumor cells demonstrated survival when encased in a biocompatible polymer membrane and implanted into the abdominal cavity of chick embryos. Several decades later, Chick et al., demonstrated that pancreatic beta cells from neonatal rats, cultured on synthetic capillaries and perfused with medium, released insulin in response to changes in glucose concentration.

In the early 1980s, Burke et al., successfully created artificial skin with fibroblasts seeded onto collagen scaffolds for the treatment of extensive burn injury. Clinically, this is still being utilized today.

Efforts are now being undertaken for engineering a variety of tissue and organ types with an emphasis on the application of stem cells.

Ultimately, the goal of tissue engineering is to regenerate tissues and restore organ function through cell implantation and matrix incorporation into the patient.

Current approaches to tissue engineering can be stratified into substitutive, histioconductive, and histioinductive. Substitutive approaches (ex vivo) are essentially whole organ replacement, whereas histioconductive approaches (ex vivo) involve the replacement of missing or damaged parts of an organ tissue with ex-vivo constructs.

In contrast, histioinductive approaches facilitate self-repair and may involve gene therapy using DNA delivery via plasmid vectors or growth factors. A number of criteria must be satisfied in order to achieve effective, long-lasting repair of damaged tissues.

- An adequate number of cells must be produced to fill the defect.
- Cells must be able to differentiate into desired phenotypes.
- Cells must adopt appropriate three-dimensional structural support/scaffold and produce ECM.
- Produced cells must be structurally and mechanically compliant with the native cell.
- Cells must successfully be able to integrate with native cells and overcome the risk of immun-ological rejection.

- There should be minimal associated biological risks.

The source of cells utilized in tissue engineering can be autologous (from the patient), allogenic (from a human donor but not immunologically identical), or xenogenic (from a different species donor).

Autologous cells represent an excellent source for use in tissue engineering because of the low association with immune complications. Autologous cells are however not cost effective and batch controlled for universal clinical use. In contrast, allogenic cells offer advantages over autologous cells in terms of uniformity, standardization of procedure, quality control and cost effectiveness.

Cell sources can be further delineated into mature (non-stem) cells, adult stem cells or somatic stem cells, embryonic stem cells (ESCs), and totipotent stem cells or zygotes.

The utility of mature cells is restricted because of its low proliferative and differentiating potential. Adult stem cells are resident stem cells found in specific niches or tissue compartments and are important in maintaining the integrity of tissues like skin, bone and blood.

They are undifferentiated cells that can be programmed to differentiate into specific tissue types. At least 20 major categories of somatic stem cells have been identified in mammals.

They can be found in bone marrow, blood, cornea and retina of the eye, dental pulp, liver, skin, GI tract, and pancreas. ESCs are derived from the inner cell mass of the pre-implantation blastocyst. They are undifferentiated, immature cells that are capable of unlimited self-renewal with the ability to differentiate.

Stem Cells (3 types)

Rapid advances are being made in stem cell research with a focus on their therapeutic potential for regenerative medicine and other biomedical applications. In combination with tissue engineering, stem cells hold a number of promises in further advancing contemporary medicine.

Traditionally, adult stem cells were believed to form a

small number of cells restricted to a particular germ layer origin; however, some evidence now indicates that adult stem cells isolated from various tissues have greater plasticity than previously thought.

Several researchers have attributed this apparent plasticity of adult stem cells to developmental signals-mediated lifferentiation. For example, Azizi et al., demonstrated that marrow stromal stem cells transplanted into the rat brain can acquire a neural phenotype.

Neural and muscle-derived stem cells isolated from cloned mice were able to differentiate into all cell types of hematopoietic lineage. A year later Clark et al., reported that murine neural stem cells could differentiate into all three germ layers with appropriate culture modifications.

In 2000 Lagasse et al., demonstrated that the injection of murine bone marrow stem cells into a mutant mouse line with progressive liver failure were able to form hepatocytes and restore liver function.

Recently, however, others reported cell fusion, rather than signal-mediated differentiation, as a possible mechanism for transdifferentiation.

For example, Alvarez- Dolado et al., demonstrated the evidence for cell fusion of bone marrow-derived cells with neurons and cardiomyocytes and the lack of trans-differentiation without cell fusion in these tissues, suggesting that cell fusion may be necessary for the observed plasticity of adult stem cells.

This finding is not without controversies. Later, others demonstrated that stem-cell plasticity is a true characteristic of neural stem cells, suggesting that transdifferentiation can be accomplished without cell fusion. In summary, the concept of adult stem cell plasticity is new, and the phenomenon is not thoroughly understood. The ESC is defined by its origin, that of a blastocyst before uterine implantation.

Three unique characteristics define primate ESCs:

- They are derived from the four to five day-old embryos that have been produced in an in vitro fertilization clinic.

- They possess a prolonged undifferentiated quality that is capable of self-renewal.
- They demonstrate stable developmental potential to form the derivatives of all three embryonic germ layers ('pluripotent') even after prolonged culture.

Pluripotent stem cells have been derived in mice and primates from pre-implantation embryos and bone marrow stroma. They can be induced to differentiate into all cell types and are able to colonize tissues of interest after transplantation. Human ESCs are mostly obtained from discarded embryos generated in fertility clinics.

In the presence of leukemia inhibitory factor (LIF) or embryonic fibroblast feeder layers, ESCs can be maintained and expanded in an undifferentiated, pluripotent state in vitro almost indefinitely. Upon withdrawal of LIF, ESCs spontaneously differentiate to form distinct embroid bodies, which contain differentiating cells of ectodermal, endodermal and mesodermal lineage. Much of the development in manipulating ESCs have been based on mouse ESC models, and there are numerous examples of improved tissue function with this usage.

Human ESCs, like no other ESCs, differentiate into somatic cell types that make up the human body. The potential benefits to health care are enormous, ranging from generating neurons for treating patients to Parkinson's disease to learning about the molecular processes that drive tumor development.

Current Progress

The current utilization of stem cells in tissue includes skin, blood vessels, cartilage, heart tissue, liver, pancreas and neural tissue. Skin defects are primarily treated by the use of epidermal and dermal constructs. Dermal fibroblasts are obtained from neonatal foreskin, expanded in vitro, seeded onto a scaffold of polylactic or polyglycolic acid before being cultured in a bioreactor system to generate a dermal layer. A bilaminate construct is produced by coating the dermal layer with multiple layers of keratinocytes. The long-term viability of these skin grafts depends on the population of the engrafted skin stem cells. Various techniques have been utilized to promote neovascularization. More recently, cell-based therapies known as endothelial progenitor cells or angioblasts have been utilized.

The lack of vascularity renders cartilage less susceptible to ischemic insult than tissues with greater dependency on direct vascularity. Autologous cell products have been used effectively for cartilage repair. Alternatively, both adult stem cells and ESCs are being investigated for their potential to differentiate into the appropriate cells for repair of damaged bone, cartilage, and tendon.

The complexity and specialized conducting infrastructure of the heart have posed a challenge to duplicate in engineering systems. Furthermore, such effort is further limited by a low proliferative potential of cardiomyocytes. Available engineered heart products are biocompatible, non-biodegradable mechanical. These devices are ineffective for longterm replacement.

Nonetheless, the possibility of development of an engineered heart is exemplified by the successful manufacturing of tissue-engineered valves and myocardial infarct scar remodeling. Combining the present engineering efforts with the potential clinical applications of stem cells may accelerate the rate of the development of these engineered products. Kehat et al., first described the generation of a reproducible spontaneous cardiomyocyte differentiating system from human ESCs. Injection of bone marrow cells in the contracting cardiac wall bordering myocardial infarcts has been demonstrated to regenerate myocardium. The development of human ESC lines and their ability to differentiate to cardiomyocytes holds great promise.

Nevertheless, bridging the gap between the theoretical potential of this unique system and any actual clinical applications will require much work. Presently, liver transplantation is the only successful treatment modality for end-stage liver failure. The limited availability of donor livers has turned our attention to alternative sources including tissue engineering.

A successful construct employs the use of highly porous biodegradable discs to deliver hepatocytes. This approach is limited by the low proliferative potential of hepatocytes and the suboptimal cell survival from the poor access to vascularity for nutrient provisions and removal of metabolic waste. Several types of stem cells, such as human adult hematopoietic and mesenchymal stem cells, have been demonstrated to differentiate into hepatocytes under the appropriate environment, and hence may serve as a unlimited cell source for therap.

Diabetes may be potentially treated by generating insulin-producing cells from stem cells and grafting them into the pancreas in diabetic patients. A much better understanding is necessary for the mechanisms that govern the expansion and differentiation into beta cells. Clinical islet cell transplantation for the treatment of type 1 diabetes has been introduced with enthusiasm.

The procedure is considered minimally invasive and potentially offers the possibility of being performed under donor-specific tolerant conditions. Recently, multipotent precursor cells from the adult mouse pancreas have been identified, and may be a promising candidate for cellbased therapeutic strategies.

Unlike isolated beta cells which release insulin in a monophasic, all-or-none manner without any fine-tuning for intermediate concentrations of blood glucose, pancreatic precursor cells may be cultured to produce all the cells of the islet cluster in order to generate a population of cells that will be able to coordinate the release of the appropriate amount of insulin to the physiologically relevant concentrations of glucose in the blood.

The potential of ESCs in the treatment of brain diseases is also under active investigation. Ma et al., recently demonstrated the first CNS stem cell-derived functional synapse and neuronal network formation on a three-dimensional collagen gel.

Mesenchymal stem cells are ideal for replacing the dead neural cells due to their strong proliferative capacity, easy acquisition, and considerable tolerance of genetic modifications.

Following the transduction of brain-derived neurotrophic factor (BDNF) via recombinant retroviral vectors into the human mesenchymal stem cells, nearly all cells were demonstrated to express BDNF, and the quantity of BDNF in the culture medium was increased by approximately 20,000-fold. Another area under investigation is genetic manipulation of adult stem cells as a potential therapeutic strategy to surmount the problem of organ shortage.

Oncology studies have identified numerous genes involved in controlling cell growth. By introducing growth-promoting genes, many stem cells can be immortalized and gain the ability to differentiate.

For example, immortalized human hepatocyte stem cells grafted into rats with acute liver failure can keep the animals alive. There is however some concern regarding oncogenic risks with using growth-promoting genes. This problem may be unlikely by introducing three levels of safeguards:

- Deletion of the transferred oncogene by site-specific recombination followed by different selection,
- Allogenic transplantation requiring temporary immunosuppression,
- Introduction of a "cellsuicide" gene that can be activated by administering a drug.

While the selection of appropriate cell source remains a key to the successful creation of stable, complex three-dimensional constructs, equally important is the design of robust scaffolds and ECM that mimics a native physiologic environment and a capillary network in a three-dimensional structure. In collaboration with Draper Laboratories, Shieh et al., have developed a computational model of vascular circulation.

Limitations

One of the major challenges of the field is to design an appropriate capillary network to allow gas exchange, provide nutrition, and remove metabolic waste from the implanted constructs. It will be necessary to create a tubular system with a minimum diameter of 10 ìm to allow vascular ingrowth, especially important for multilayered solid organ constructs with a volume greater than 2 to 3 cubic mm. An alternative is to implant a vascularized construct in an ectopic, well-vascularized location, such as omentum.

Special attention will be given to the control of the local milieu of each cell type as different cell types often require different culturing environments, making it difficult to design a multilayered organ system construct. A challenge will be to create a scaffold capable of supporting various cell types. The appropriate expression sequence, dosing, and duration for growth factors also needs to be defined if we are to maximize ESCs utility in tissue growth and maintenance in the pluripotent state.

Ethical/technical Considerations

There are three main research programmes in stem cell research.

- Research on adult stem cells.
- Research on ESCs from discarded embryos that are produced via in vitro fertilization.
- Research on ESCs obtained via therapeutic cloning.

There is little controversy surrounding human adult stem cells. On the other hand, human ESCs face great controversies, the extent of which is dependent on the type of research programmes.

Various autologous adult stem cells, including mesenchymal, hematopoietic, neural, muscle, and hepatic stem cells, are actively being investigated. They are more likely immunocompatible and are not associated with ethical concerns unlike their ESC counterparts. Challenges remain in regard to optimization of isolation techniques that minimize contamination, permanently maintain the desired cell types following differentiation, and increase the production of a large number of cells that are adequate for the construction of a tissue or organ.

Furthermore, much of the data cited have not been

reproducible. The utility of adult stem cells would effectively avoid the ethical problem of using ESCs. But much needs to be learned about adult stem cells in regard to their replication and differentiation. ESCs may offer technical advantages over other sources of somatic precursors. For example, ESCs can be generated in abundant quantities in the laboratory and can be grown in their undifferentiated state for many generations.

In contrast, researchers have had difficulty finding laboratory conditions under which some adult stem cells, especially hematopoetic stem cells from blood or bone marrow, can proliferate without becoming specialized. This technical barrier to proliferation has limited the ability of researchers to explore the capacity of certain types of adult stem cells to generate sufficient numbers of specialized cells for transplantation purposes.

Unlike adult stem cells, ESCs can be grown indefinitely in tissue culture. There is an increased risk for immune rejection because of genetic mismatch, which in turn would demand lifelong high-dose immunosuppressant treatment. Furthermore, the creation of human ESCs derived from discarded, non-transferred human embryos using an immunosurgical technique has resumed arguments that have not been resolved.

A SART-RAND study identified at least 400,000 frozen embryos in storage since the late 1970s. Only 2.8 per cent of these embryos have been designed for research. With a 15 per cent efficiency rate for generating a human ESC line from blastocysts and incorporating some losses to the freeze-thaw process, only approximately 275 human ESC lines can be created from the excess. The pluripotential virtue of stem cells itself renders limitations: the efficiency of differentiation appears to be compromised for certain cell types, such as hepatocytes and cardiomyocytes, and the quality of differentiation is still questionable.

The use of ESCs as an important cell source for tissue engineering has great potential for the treatment of diseases, but there are alternatives, such as the use of adult stem cells and ESCs from therapeutic cloning under active investigation.

Worldwide legislation reflects the dilemma, ranging from the more permissive to the nominally more restrictive (the United States—only minimal federal financial funds granted to research that meets the strict NIH guidelines, but stem cell research is legal without any restriction in the private sector) to the truly restrictive (Germany, Austria, Ireland, France—laws on reproductive medicine ban the extraction from stem cells from a human embryo).

The host immune response to allogenic embryonic stem cells poses a major challenge in addition to ethical and religious concerns regarding ESCs. This has led researchers to investigate creation of a universal donor cell by making histocompatible proteins on the cell surface thus reducing the cells' antigenicity.

This is the heart of nuclear transfer technology or "therapeutic cloning". However, the technology has its own shortcomings. It has a low efficiency rate which in turn requires a large number of oocytes. Its inability to control differentiation requires a better understanding of ESC growth, differentiation, and application in engineering tissues. Furthermore, therapeutic cloning protocols in vitro culture systems must be optimized before contemplating the use of this technology for cell therapy.

All three research programmes discussed here are directed at increasing our knowledge about basic cell biology, creating new therapies through stem cell culture and control of cell differentiation, and producing commercially viable stem cell products either by the direct patenting of stem cell lines or by combining stem cell technologies with genetic engineering or other patentable interventions.

The primary discussion on stem cells is concerned with the ethical issues raised by each of these programmes and with whether these issues should influence regulatory decisions regarding public financial support for research.

Proposition Passage in California

The California Stem Cell Research and Cures Initiative, recently passed by statewide ballot in November 2004. As a result, the California Institute for Regenerative Medicine (CIRM or Institute) was created in order to provide breakthrough cures for devastating diseases. These diseases include, but are not limited to, diabetes, heart diseases, cancer, osteoporosis, multiple sclerosis, Alzheimer's disease, Parkinson's disease, Lou Gerig's disease and spinal cord injuries by utilizing tax-free state bonds to fund stem cell research.

The institute would ensure that strict ethics codes will be adopted and enforced, including utilizing guidelines developed by the National Institutes of Health. Mirroring the existing state law, the institute would strictly prohibit human reproductive cloning.

In addition to finding cures for various diseases, the goal of the institute is also to reduce the rapidly rising healthcare costs in California and boost its economy. Proposition is designed not to burden taxpayers, authorizing tax-free state bonds that will provide an average of $295 million per year over ten years to support stem cell research in California. It assumes that a large share of Health care costs is caused by diseases and injuries that can potentially be cured with stem cell therapies.

Opponents to Proposition have argued that the bill places an unnecessary and unwise financial burden on an already-insufficient state budget. According to some estimates, state bond debt, which amounted to $33 billion in May of 2004, might rise to more than $50 billion by July of 2005, largely due to the costs of Proposition.

In addition, the initiative has yet to assure the Californian taxpayers whether the state will actually receive any royalties from discoveries paid for by state dollars. Questions about significant conflicts of interest on the Independent Citizen's Oversight Committee, (the "board of directors" of CIRM) and even about the scientific integrity of the stem-cell lines available for research have been raised.

In spite of this opposition, many defend the economic merits of Proposition. As no payments would be demanded within the first five years of the bill's implementation, the state would remain free from financial stress and allowed time to balance its budget. No tax increases would accompany the legislation.

The potential effects that the passage of California's Proposition would have on other states and the nation as a whole are not to be underestimated. New Jersey, the nation's first state to create a publicly funded institute for stem cell research and the second state to pass legislation legalizing the research, has found it necessary to review its policy and is considering an increase in their research grants.

Other states, including Massachusetts, Maryland, Wisconsin and Connecticut, are also jumping in with proposals to fund research cloning.

These changes are happening against the recent passage of the "declaration" of banning all forms of human cloning by the United Nations. A recent Swiss law that mirrors laws in most European countries, allows medical research on stem cells taken from human embryos but bars cloning. Britain and Belgium allow embryo creation for research. Britain has an independent embryo authority to scrutinize and monitor all projects and has issued only two cloning licenses since 2001.

What has made the United States such a fertile ground for expanding embryo research is not its liberal laws but the lack of them. Congress has tried but failed to pass legislation largely because of irreconcilable differences over when life begins.

Both Presidents Bill Clinton and George W. Bush banned the National Institutes of Health on ethical grounds from funding the creation of human research embryos, although their orders apply only to federally funded work. In 2001, Bush additionally narrowed NIH research to embryonic stem cells already harvested from spare IVF embryos. These restrictions in turn triggered an influx of private and state funding, and the dramatic Proposition.

Future of Tissue Engineering with Stem cell Research

Although much progress has been made in the field of tissue engineering, further work toward organ and tissue replacement is necessary. The optimal cell source, scaffold design, and in vitro bioreactors, the use and development of microfabrication technology to create vascularized tissues and organs are still being investigated.

The search for and use of an appropriate multipotent or pluripotent stem cell in tissue engineering is an emerging concept.

Certainly, many areas of stem cell research and their potential clinical applications are associated with controversies; therefore, it is important to address the ethical, legal, and social issues early. Many technical questions are yet to be answered and require close interdisciplinary collaborations of surgeons, engineers, chemists, and biologists, with the ultimate goal of functional tissue restoration.

As more scientific knowledge will be gained from stem cell research, hopefully, some of the current ethical and technical concerns will be answered or removed in the future.

Chapter 7

Manipulating Cells and Viruses in Culture

Understanding modern molecular cell biology requires a familiarity with the most commonly used biological materials and knowledge of the latest experimental techniques. In this chapter, we describe the primary techniques used to culture and manipulate certain cells, viruses, and experimental organisms.

Biologists choose to study these cells and organisms because they provide experimentally favorable examples of important molecular events or processes such as the control of gene activity and cell replication, formation of organelles, secretion of proteins, and differentiation of cells.

In the first part of the chapter, we discuss the culture of microbial cells, animal cells, and hybrid cells in vitro and give some examples of their experimental uses. Many studies involve cells or organisms with mutations affecting a specific biochemical or developmental pathway. In such studies, an important criterion for using a particular cell type or organism is the ease with which mutants can be isolated and characterized and the mutated genes identified and cloned. Because viruses contain a small number of genes (4 to ≈100 in different types of viruses), viral mutants commonly are used to study the action of a limited set of genes designed to carry out restricted, specific molecular tasks. To provide a basis for understanding such studies, we consider the basic properties of viruses and their uses in various types of research in the latter part of this chapter.

GROWTH OF MICROORGANISMS IN CULTURE

Cultured cells have several advantages over intact organisms for research on fundamental aspects of cell biology. First, most animal and plant tissues comprise a variety of different types of cells, whereas cells of a specific type can be grown in culture; thus cultured cells are more homogeneous in their properties than in vivo cells in tissues.

Second, experimental conditions can be controlled much more rigorously in culture than in an organism. By manipulating the growth conditions, for example, one can measure the effects of specific chemicals and growth factors on a particular cell type in culture.

A third reason for preferring cultured cells is that in many cases a single cell can be readily grown into a colony, a process called *cell cloning,* or simply *cloning.*

The resulting strain of cells, which is genetically homogeneous, is called a clone. This simple technique, which is commonly used with many bacteria, yeasts, and mammalian cell types, makes it easy to isolate genetically distinct clones of cells.

MICROORGANISMS CAN BE GROWN IN MINIMAL MEDIUM

Among the advantages of using microorganisms such as the bacterium *Escherichia coli* and the yeast *Saccharomyces cerevisiae* are their rapid growth rate and simple nutritional requirements, which can be met with a minimal medium. A minimal medium for such microorganisms can contain glucose as the sole source of carbon; metabolism of glucose to smaller molecules (e.g., CO_2, ethanol, or acetic acid) can generate the ATP necessary for energy-requiring activities of the cells.

The sole nitrogen source in a minimal medium can be ammonium (NH_4^+), from which the cells can synthesize all the necessary amino acids and other nitrogen-containing metabolites. Salts and trace elements are the only other components of a minimal medium.

Many prokaryotes (i.e., bacteria) and single-celled eukaryotes such as yeast, both of which grow in nature as single cells, are easily grown in culture dishes — usually on top of *agar*, a semisolid base of plant polysaccharides. The agar is first dissolved in a heated nutrient medium, and the solution is poured into petri dishes; as it cools, the solution solidifies. A dilute suspension of cells then is dispersed on top of the agar; in time each cell grows into a discrete colony. Since the cells in a colony all derive from a single cell, they form a clone and have identical genomes (DNA). All the cells in a clone generally express the same set of genes and contain the same enzymes and other constituents in similar proportions.

The division time for *E. coli* and similar microorganisms ranges from 20 minutes to 1 hour. Thus a single *E. coli* cell, which divides approximately every 30 minutes, can grow into a colony containing $10^7 - 10^8$ cells in 12 hours ($2^{24} = 1.7 \times 10^7$). Because yeasts, unlike most eukaryotic organisms, grow as single cells and can be grown in a simple defined medium similarly to bacteria, they are a popular choice for studies of eukaryotic cell function. The entire life cycle of yeasts can be studied in culture, and colonies can be grown from a single vegetative cell (a growing cell) or from a single spore (a dormant cell).

Studies with yeasts have provided valuable information on such subjects as the cellular mechanisms for controlling DNA synthesis, genetic recombination, and the gene products that regulate the cycle of events leading to replication and segregation of chromosomes and ultimately cell division.

MUTANT STRAINS OF BACTERIA

If a random mutation in a microorganism cultured, or plated, on a minimal agar medium renders cells unable to synthesize a necessary metabolite, and this metabolite is not supplied in the medium (e.g., the amino acid arginine), then the cells will die. But if the cells are grown on a rich medium that supplies the metabolite whose synthesis is prevented by a mutation, then the mutant cell can survive and grow into a colony in which all the cells have the same mutation.

A mutant colony, say, an Arg^+ colony, can be detected by *replica plating,* a technique developed by Joshua and Esther Lederberg in the 1940s. The initial plate of multiple colonies growing on rich medium is called the *master plate.* A circular, velvet-covered stamp equal in size to the master plate is pressed onto it, so that the stamp picks up cells from each colony.

The stamp is then used to deposit cells onto a second plate containing minimal agar medium without arginine (Arg^+ medium) and also onto a third plate containing minimal agar medium with arginine (Arg^+ medium). Cells that can synthesize arginine grow into colonies on both the Arg^+ and Arg^+ plates.

However, a clone of cells on the master plate with a mutation in a gene required for arginine synthesis cannot grow on the Arg^+ medium, resulting in a space on the Arg^+ plate corresponding to the position of this clone on the master plate and Arg^+ plate. This clone of cells, thus identified as being defective in the synthesis of arginine, then can be isolated from the master plate.

Repeated replica plating of a master plate with paired minimal-medium plates containing or lacking different specific nutrients (e.g., amino acids, nucleic acid bases, vitamins) can identify clones of cells with mutations in the genes required for synthesis of different biological molecules.

This approach has been used to isolate thousands of different bacterial and yeast nutritional auxotrophs; these are strains that are unable to grow in the absence of a specific nutrient.

Under standard culture conditions, random mutations occur in *E. coli* genes at a frequency of 1 in $10^6 - 10^8$ base pairs per generation. Thus, by the time a single cell has divided to produce a few hundred or thousand cells, some individual genes will differ among cells in the clone.

To maintain the genetic homogeneity of a clone, it must be recloned frequently under selective conditions that permit only cells with a particular set of desired characteristics to survive and grow.

For example, to maintain an Arg^+ clone, isolated as described above, the cells are replated on minimal medium plus arginine; only cells that can synthesize all the metabolites required for growth except arginine will grow on this selective medium.

GROWTH OF ANIMAL CELLS IN CULTURE

Animal cells are more difficult to culture than microorganisms because they require many more nutrients and typically grow only when attached to specially coated surfaces. Despite these difficulties, various types of animal cells, including both undifferentiated and differentiated ones, can be cultured successfully.

Culture of Animal Cells

Nine amino acids, referred to as the *essential amino acids,* cannot be synthesized by adult vertebrate animals and thus must be obtained from their diet. Animal cells grown in culture also must be supplied with these nine amino acids, namely, histidine, isoleucine, leucine, lysine, methionine, phenylalanine, threonine, tryptophan, and valine.

In addition, most cultured cells require cysteine, glutamine, and tyrosine. In the intact animal, these three amino acids are synthesized by specialized cells; for example, liver cells make tyrosine from phenylalanine, and both liver and kidney cells can make glutamine.

Animal cells both within the organism and in culture can synthesize the 8 remaining amino acids; thus these amino acids need not be present in the diet or culture medium. The other essential components of a medium for culturing animal cells are vitamins, which the cells cannot make at all or in adequate amounts; various salts; glucose; and *serum,* the noncellular part of the blood. Serum, a mixture of hundreds of proteins, contains various factors needed for proliferation of cells in culture. For example, it contains insulin, a hormone required for growth of many cultured vertebrate cells, and transferrin, an iron-transporting protein essential for incorporation of iron by cells in culture. Although many animal cells can grow in a serum-containing medium, such as Eagle's medium, certain cell types require specific protein growth factors that are not present in serum.

For instance, precursors of red blood cells require the hormone erythropoietin, and T lymphocytes of the immune system require interleukin 2 (IL-2). These factors bind to receptor proteins that span the plasma membrane, signaling the cells to increase in size and mass and undergo cell division. A few mammalian cell types can be grown in a completely defined, serum-free medium supplemented with trace minerals, specific protein growth factors, and other components.

Cultured Animal Cells Grow Only on Special Solid Surfaces

Within the tissues of intact animals, most cells tightly contact and interact specifically with other cells via various cellular junctions. The cells also contact the extracellular matrix, a complex network of secreted proteins and carbohydrates that fills the spaces between cells.

The matrix, whose constituents are secreted by cells themselves, helps bind the cells in tissues together; it also provides a lattice through which cells can move, particularly during the early stages of animal differentiation. The extracellular matrices in various animal tissues consist of several common components: fibrous collagen proteins; hyaluronan (or hyaluronic acid), a large mucopolysaccharide; and covalently linked polysaccharides and proteins in the form of proteoglycans (mostly carbohydrate) and glycoproteins (mostly protein).

However, the exact composition of the matrix in different tissues varies, reflecting the specialized function of a tissue. In connective tissue, for example, the major protein of the extracellular matrix is a type of collagen that forms insoluble fibers with a very high tensile strength.

Fibroblasts, the principal cell type in connective tissue, secrete this type of collagen as well as the other matrix components. Receptor proteins in the plasma membrane of cells bind various matrix elements, imparting strength and rigidity to tissues.

The tendency of animal cells in vivo to interact with one another and with the surrounding extracellular matrix is mimicked in their growth in culture. Unlike bacterial and yeast cells, which can be grown in suspension, most cultured animal cells require a surface to grow on. Many types of cells can adhere to and grow on glass, or on specially treated plastics with negatively charged groups on the surface (e.g., SO_3^{2-}).

The cultured cells secrete collagens and other matrix components; these bind to the culture surface and function as a bridge between it and the cells. Cells cultured from single cells on a glass or a plastic dish form visible colonies in 10 – 14 days. Some tumor cells can be grown in suspension, a considerable experimental advantage because equivalent samples are easier to obtain from suspension cultures than from colonies grown in a dish.

PRIMARY CELL CULTURES ARE USEFUL

Normal animal tissues (e.g., skin, kidney, liver) or whole embryos commonly are used to establish *primary cell cultures.* To prepare tissue cells for culture (or to remove adherent cells from a culture dish for biochemical studies), trypsin or another protease is used to destroy the proteins in the junctions that normally interconnect cells. For many years, most cell types were difficult, if not impossible, to culture.

But the identification and preparation of various protein growth factors that stimulate the replication of specific cell types, as well as other recent modifications in culture methods, now permit experimenters to grow various types of specialized cells. Many studies with vertebrate cells, however, still are performed with those few cell types that grow most readily in culture. These are not cells of a defined type; rather, they represent whatever grows when a tissue or an embryo is placed in culture.

The cell type that usually predominates in such cultures is called a fibroblast because it secretes the types of proteins associated with fibroblasts in fibrous connective tissue of animals. Cultured fibroblasts have the morphology of tissue fibroblasts, but they retain the ability to differentiate into other cell types; thus they are not as differentiated as tissue fibroblasts.

Some studies are conducted with primary cultures of epithelial cells. In general, external and internal surfaces of tissues and organs are covered by a layer of epithelial cells called an epithelium. These highly differentiated cells are said to be *polarized* because the plasma membrane is organized into at least two discrete regions.

For example, the epithelial cells that line the intestine form a simple columnar epithelium. That portion of the plasma membrane facing the lumen of the intestine, the *apical* surface, is specialized for absorption; the rest of the plasma membrane, the *basolateral* surface, mediates transport of nutrients from the cell to the blood and forms junctions with adjacent cells and the underlying extracellular matrix called the basal lamina.

Certain cells cultured from blood, spleen, or bone marrow adhere poorly, if at all, to a culture dish but nonetheless grow well. In the body, such nonadherent cells are held in suspension (in the blood), or they are loosely adherent (in the bone marrow and spleen). Because these cells often come from immature stages in the development of differentiated blood cells, they are very useful for studying normal blood cell differentiation and the abnormal development of leukemias.

When cells are removed from an embryo or an adult animal, most of the adherent ones grow continuously in culture for only a limited time before they spontaneously cease growing. Such a culture eventually dies out after many cell doublings, even if it is provided with fresh supplies of all the known nutrients that cells need to grow, including serum.

For instance, when human fetal cells are explanted into cell culture, the majority of cells die within a relatively short time; "fibroblasts," although also destined to die, proliferate for a while and soon become the predominant cell type.

They divide about 50 times before they cease growth. Starting with 10^6 cells, 50 doublings can produce $10^6 \times 2^{50}$, or more than 10^{20} cells, which is equivalent to the weight of about

10^5 people. Thus, even though its lifetime is limited, a single culture, if carefully maintained, can be studied through many generations. Such a lineage of cells originating from one initial primary culture is called a cell strain.

Transformed Cells Can Grow Indefinitely in Culture

To be able to clone individual cells, modify cell Behaviour, or select mutants, biologists often want to maintain cell cultures for many more than 100 doublings. This is possible with cells derived from some tumors and with rare cells that arise spontaneously because they have undergone genetic changes that endow them with the ability to grow indefinitely.

The genetic changes that allow these cells to grow indefinitely are collectively called *oncogenic* transformation, and the cells are said to be *oncogenically transformed,* or simply *transformed.* A culture of cells with an indefinite life span is considered immortal; such a culture is called a cell line to distinguish it from an impermanent cell *strain.*

The ability of cultured cells to grow indefinitely or their tendency to be transformed varies depending on the animal species from which the cells originate. Normal chicken cells rarely are transformed and die out after only a few doublings; even tumor cells from chickens almost never exhibit immortality. Among human cells, only tumor cells grow indefinitely. The HeLa cell, the first human cell type to be grown in culture, was originally obtained in 1952 from a malignant tumor (carcinoma) of the uterine cervix. This cell line has been invaluable for research on human cells.

In contrast to human and chicken cells, cultures of embryonic adherent cells from rodents routinely give rise to cell lines. When adherent rodent cells are first explanted, they grow well, but after a number of serial replatings they lose growth potential and the culture goes into crisis. During this period most of the cells die, but often a rapidly dividing variant cell arises spontaneously and takes over the culture.

A cell line derived from such a variant will grow forever if it is provided with the necessary nutrients. Cells in spontaneously established rodent cell lines and in cell lines derived from tumors often have abnormal chromosomes. In addition, their chromosome number usually is greater than that of the normal cell from which they arose, and it continually expands and contracts in culture. Such cells are said to be *aneuploid* (i.e., have an inappropriate number of chromosomes) and are obviously mutants.

Although most cell lines are undifferentiated, some can carry out many of the functions characteristic of the normal differentiated cells from which they are derived. One example is certain hepatoma cell lines (e.g., HepG2) that synthesize most of the serum proteins made by normal hepatocytes (the major cell type in the liver) from which they are derived.

These highly differentiated hepatoma cells are often studied as models of normal hepatocytes. Cultured *myoblasts* (muscle precursor cells) are another example of transformed cells that continue to perform many functions of a specialized, differentiated cell. When grown in culture, transformed myoblasts can be induced to fuse to form myotubes. These resemble differentiated multinucleated muscle cells and synthesize many of, if not all, the specialized proteins associated with contraction.

Certain lines of epithelial cells also have been cultured successfully. One such line, Madin-Darby canine kidney (MDCK) cells, forms a continuous sheet of polarized epithelial cells one cell thick that exhibits many of the properties of the normal canine kidney epithelium from which it was derived. This type of preparation has proved valuable as a model for studying the functions of epithelial cells.

FUSION OF CULTURED ANIMAL CELLS

Cultured animal cells infrequently undergo cell fusion spontaneously. The fusion rate, however, increases greatly in the presence of certain viruses that have a lipoprotein envelope similar to the plasma membrane of animal cells. A mutant viral glycoprotein in the envelope promotes cell fusion the mechanism of this effect is discussed. Cell fusion also is promoted by polyethylene glycol, which causes the plasma membranes of adjacent cells to adhere to each other and to fuse. As most fused animal cells undergo cell division, the nuclei eventually fuse, producing viable cells with a single nucleus that contains chromosomes from both "parents." The fusion of two cells that are genetically different yields a hybrid cell called a heterokaryon.

Because some somatic cells from animals can be cultured from single cells in a well-defined medium, it is possible to select for genetically distinct cultured animal cells, just as is done with bacterial and yeast cells. Moreover, during mitosis the chromosomes in an animal cell are large and highly visible after staining, making it easy to distinguish individual chromosomes. Genetic studies of cultured animal cells are called *somatic-cell genetics* to distinguish them from *classical genetics,* which deals with whole organisms derived from germ cells (sperm and eggs). Cultured cells from different mammals can be fused to produce interspecific hybrids, which have been widely used in somatic-cell genetics. For instance, hybrids can be prepared from human cells and mutant mouse cells that lack an enzyme required for synthesis of a particular essential metabolite. As the human-mouse hybrid cells grow and divide, they gradually lose human chromosomes in random order, but retain the mouse chromosomes.

In a medium that can support growth of both the human cells and mutant mouse cells, the hybrids eventually lose all human chromosomes. However, in a medium lacking the essential metabolite that the mouse cells cannot produce, the one human chromosome that contains the gene encoding the needed enzyme will be retained, because any hybrid cells that lose it following mitosis will die. All other human chromosomes eventually are lost. By using different mutant mouse cells and media in which they cannot grow, researchers have prepared various panels of hybrid cell lines. Each cell line in a panel contains either a single human chromosome or a small number of human chromosomes, and a full set of mouse chromosomes.

Because each chromosome can be identified visually under a light microscope, such hybrid cells provide a means for assigning, or "mapping," individual genes to specific chromosomes. For example, suppose a hybrid cell line is shown microscopically to contain a particular human chromosome. That hybrid cell line can then be tested biochemically for the presence of various human enzymes, exposed to specific antibodies to detect human surface antigens, or subjected to DNA hybridization and cloning techniques to locate particular human DNA sequences. The genes encoding a human protein or containing a human DNA sequence detected in such tests must be located on the particular human chromosome carried by the cell line being tested. Panels of hybrids between normal mouse and mutant hamster cells also have been established; in these hybrid cells, the majority of mouse chromosomes are lost, allowing mouse genes to be mapped to specific mouse chromosomes.

HYBRID CELLS

One metabolic pathway has been particularly useful in cell-fusion experiments. Most animal cells can synthesize the purine and pyrimidine nucleotides de novo from simpler carbon and nitrogen compounds, rather than from already formed purines and pyrimidines.

The folic acid antagonists amethopterin and aminopterin interfere with the donation of methyl and formyl groups by tetrahydrofolic acid in the early stages of de novo synthesis of glycine, purine nucleoside monopho sphates, and thymidine monophosphate.

These drugs are called *antifolates,* since they block reactions involving tetrahydrofolate, an active form of folic acid. Many cells, however, contain enzymes that can synthesize the necessary nucleotides from purine bases and thymidine if they are provided in the medium; these *salvage pathways* bypass the metabolic blocks imposed by antifolates. A number of mutant cell lines lacking the enzyme needed to catalyze one of the steps in a salvage pathway have been isolated. For example, cell lines lacking thymidine kinase (TK) can be selected because such cells are resistant to the otherwise toxic thymidine analog 5-bromodeoxyuridine. Cells containing TK convert 5-bromodeoxyuridine into 5-bromodeoxyuridine monophosphate.

This nucleoside mono- phosphate is then converted into a nucleoside triphosphate by other enzymes and is incorporated by DNA polymerase into DNA, where it exerts its toxic effects.

This pathway is blocked in cells with a *TK* mutation that prevents production of functional TK enzyme. Hence, TK″ mutants are resis- tant to the toxic effects of 5-bromodeoxyuridine.

Similarly, cells lacking the HGPRT enzyme have been selected because they are resistant to the otherwise toxic guanine analog 6-thioguanine.

As we will see next, HGPRT″ cells and TK″ cells are useful partners in cell fusions with one another or with cells that have salvage-pathway enzymes but that are differentiated and cannot grow in culture by themselves.

The medium most often used to select hybrid cells is called *HAT medium,* because it contains hypoxanthine (a purine), aminopterin, and thymidine.

Normal cells can grow in HAT medium because even though aminopterin blocks de novo synthesis of purines and TMP, the thymidine in the media is transported into the cell and converted to TMP by TK and the hypoxanthine is transported and converted into usable purines by HGPRT. On the other hand, neither TK″ nor HGPRT″ cells can grow in HAT medium because each lacks an enzyme of the salvage pathway.

However, hybrids formed by fusion of these two mutants will carry a normal *TK* gene from the HGPRT″ parent and a normal *HGPRT* gene from the TK″parent.

The hybrids thus will produce both functional salvage-pathway enzymes and grow on HAT medium. Likewise, hybrids formed by fusion of mutant cells and normal cells can grow in HAT medium.

For many types of studies involving antibodies, monoclonal antibody is preferable to polyclonal antibody. However, biochemical purification of monoclonal antibody from serum is not feasible, in part because the concentration of any given antibody is quite low. For this reason, researchers looked to culture techniques in order to obtain usable quantities of monoclonal antibody.

Because primary cultures of normal B lymphocytes do not grow indefinitely, such cultures have limited usefulness for production of monoclonal antibody.

This limitation can be avoided by fusing normal B lymphocytes with oncogenically transformed lymphocytes called *myeloma cells,* which are immortal.

Fusion of a myeloma cell with a normal antibody-producing cell from a rat or mouse spleen yields a hybrid that proliferates into a clone called a hybridoma. Like myeloma cells, hybridoma cells are immortal. Each hybridoma produces the monoclonal antibody encoded by its B-lymphocyte partner.

Many different myeloma cell lines from mice and rats have been established; from these, HGPRT″ lines have been selected based on their resistance to 6-thioguanine as described above. If such mutant myeloma cells are fused with normal B lymphocytes, any fused cells that result can grow in HAT medium, but the parental cells cannot.

Each selected hybridoma then is tested for production of the desired antibody; any clone producing that antibody then is grown in large cultures, from which a substantial quantity of pure monoclonal antibody can be obtained. Such pure antibodies are very valuable research reagents. For example, a monoclonal antibody that interacts with protein X can be used to label, and thus locate, protein X in specific cells of an organ or in specific cell fractions.

Once identified, even very scarce proteins can be isolated by affinity chromatography in columns to which the monoclonal antibody is bound.

Monoclonal antibodies also have become important diagnostic and therapeutic tools in medicine. Monoclonal antibodies that bind to and inactivate toxic proteins (toxins) secreted by bacterial pathogens are used to treat diseases caused by these pathogens.

Other monoclonal antibodies are specific for cell-surface proteins expressed by certain types of tumor cells; chemical complexes of such monoclonal antibodies with toxic drugs are being developed for cancer chemotherapy.

VIRUSES: STRUCTURE, FUNCTION, AND USES

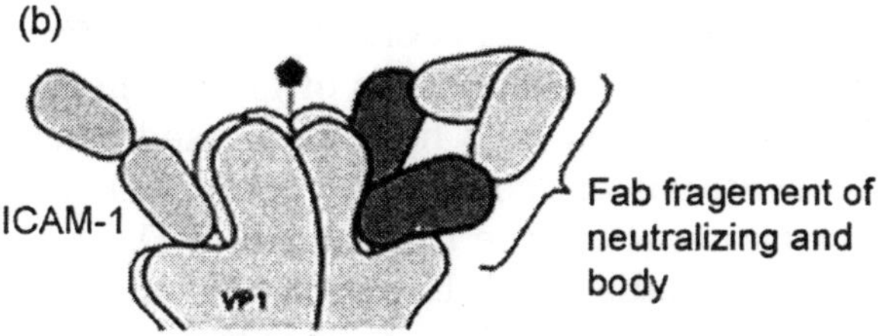

Fig. Structure of picornaviruses.

A virus is a small parasite that cannot reproduce by itself. Once it infects a susceptible cell, however, a virus can direct the cell machinery to produce more viruses. Most viruses have either RNA or DNA as their genetic material. The nucleic acid may be single- or double-stranded.

The entire infectious virus particle, called a virion, consists of the nucleic acid and an outer shell of protein. The simplest viruses contain only enough RNA or DNA to encode four proteins. The most complex can encode 100 – 200 proteins. The study of plant viruses inspired some of the first experiments in molecular biology. In 1935, Wendell Stanley purified and partly crystallized tobacco mosaic virus (TMV); other plant viruses were crystallized soon thereafter. Pure proteins had been crystallized only a short time before Stanley's work, and it was considered very surprising at the time that a replicating organism could be crystallized.

A wealth of subsequent research with bacterial viruses and animal viruses has provided detailed understanding of viral structure, and virus-infected cells have proved extremely useful as model systems for the study of basic aspects of cell biology. In many cases, DNA viruses utilize cellular enzymes for synthesis of their DNA genomes and mRNAs; all viruses utilize normal cellular ribosomes, tRNAs, and translation factors for synthesis of their proteins. Most viruses commandeer the cellular machinery for macromolecular synthesis during the late phase of infection, directing it to synthesize large amounts of a small number of viral mRNAs and proteins instead of the thousands of normal cellular macromolecules. For instance, animal cells infected by influenza or vesicular stomatitis virus synthesize only one or two types of glycoproteins, which are encoded by viral genes, whereas uninfected cells produce hundreds of glycoproteins.

Such virus-infected cells have been used extensively in studies on synthesis of cell-surface glycoproteins. Similarly, much information about the mechanism of DNA replication has come from studies with bacterial cells and animal cells infected with simple DNA viruses, since these viruses depend almost entirely on cellular proteins to replicate their DNA. Viruses also often express proteins that modify host-cell processes so as to maximize viral replication. For example, the roles of certain cellular factors in initiation of protein synthesis were revealed because viral proteins interrupt their action. Finally, when certain genes carried by cancer-causing viruses integrate into chromosomes of a normal animal cell, the normal cell can be converted to a cancer cell.

Since many viruses can infect a large number of different cell types, genetically modified viruses often are used to carry foreign DNA into a cell. This approach provides the basis for a growing list of experimental gene therapy treatments. Because of the extensive use of viruses in cell biology research and their potential as therapeutic agents, we describe the basic aspects of viral structure and function in this section.

VIRAL CAPSIDS ARE REGULAR

The nucleic acid of a virion is enclosed within a protein coat, or capsid, composed of multiple copies of one protein or a few different proteins, each of which is encoded by a single viral gene. Because of this structure, a virus is able to encode all the information for making a relatively large capsid in a small number of genes.

This efficient use of genetic information is important, since only a limited amount of RNA or DNA, and therefore a limited number of genes, can fit into a virion capsid. A capsid plus the enclosed nucleic acid is called a nucleocapsid.

Nature has found two basic ways of arranging the multiple capsid protein subunits and the viral genome into a nucleocapsid. The simpler structure is a protein helix with the RNA or DNA protected within.

Tobacco mosaic virus (TMV) is a classic example of the helical nucleocapsid. In TMV the protein subunits form broken disklike structures, like lock washers, which form the helical shell of a long rodlike virus when stacked together.

The other major structural class of viruses, called *icosahedral* or *quasi-spherical viruses,* is based on the icosahedron, a solid object built of 20 identical faces, each of which is an equilateral triangle. In the simplest type of icosahedral virion each of the 20 triangular faces is constructed of three identical capsid protein subunits, making a total of 60 subunits per capsid. At each of the 12 vertices, five subunits make contact symmetrically.

Thus all protein subunits are in *equivalent* contact with one another. Tobacco satellite necrosis virus has such a simple icosahedral structure.

However, most quasi-spherical viruses are larger, requiring the assembly of more than three subunits per face of the icosahedron. These proteins form shells whose subunits are in *quasi-equivalent* contact. Here, the proteins at the icosahedral vertices remain arranged in a fivefold symmetry, but additional subunits cover the surfaces between in a pattern of sixfold symmetry.

The atomic structures of a number of icosahedral viruses have been determined by x-ray crystallography. The first three such viruses to be analyzed — tomato bushy stunt virus, poliovirus, and rhinovirus (the common cold virus) — exhibit a remarkably similar design, in terms of the rules of icosahedral symmetry as well as in the details of their surface proteins.

In each virus, at atomic resolution, clefts ("canyons") are observed encircling each of the vertices of the icosahedral structure. Interaction of these clefts with cell-surface receptors attaches the virus to a host cell, the first step in viral infection.

Neutralizing antibodies specific for a particular virus also interact with these clefts, thereby inhibiting attachment of the virus to the host cell. In some viruses, the symmetrically arranged nucleocapsid is covered by an external membrane, or *envelope,* which consists mainly of a phospholipid bilayer but also contains one or two types of virus-encoded glycoproteins. The phospholipids in the viral envelope are similar to those in the plasma membrane of an infected host cell. The viral envelope is, in fact, derived by budding from that membrane, but contains mainly viral glycoproteins.

The components of simple viruses such as TMV, which consists of a single RNA molecule and one protein species, undergo self-assembly if they are mixed in solution. More complex viruses containing a dozen or more protein species do not spontaneously assemble in vitro.

The multiple components of such viruses assemble within infected cells in stages, first into subviral particles and then into completed virions. The genomes of these complex viruses encode proteins that assist in the assembly of the virion, but the assembly proteins are not themselves components of the completed virion.

MOST VIRAL HOST RANGES ARE NARROW

The fact that the host range — the group of cell types that a virus can infect — is generally restricted serves as a basis for classifying viruses. A virus that infects only bacteria is called a bacteriophage, or simply a phage. Viruses that infect animal or plant cells are referred to generally as *animal viruses* or *plant viruses.*

A few viruses can grow in both plants and the insects that feed on them. The highly mobile insects serve as vectors for transferring such viruses between susceptible plant hosts. An example is potato yellow dwarf virus, which can grow in leafhoppers (insects that feed on potato plant leaves) as well as in potato plants. Wide host ranges are characteristic of some strictly animal viruses, such as vesicular stomatitis virus, which grows in insects and in many different types of mammalian cells.

Most animal viruses, however, do not cross phyla, and some (e.g., poliovirus) infect only closely related species such as primates. The host-cell range of some animal viruses is further restricted to a limited number of cell types because only these cells have appropriate surface receptors to which the virions can attach.

VIRUSES CAN BE CLONED

The number of infectious viral particles in a sample can be quantified by a plaque assay. This assay is performed by culturing a dilute sample of viral particles on a plate covered with host cells and then counting the number of local lesions, called *plaques,* that develop.

A plaque develops on the plate wherever a single virion initially infects a single cell. The virus replicates in this initial host cell and then lyses the cell, releasing many progeny virions that infect the neighboring cells on the plate. After a few such cycles of infection, enough cells are lysed to produce a visible plaque in the layer of remaining uninfected cells.

Since all the progeny virions in a plaque are derived from a single parental virus, they constitute a virus clone. This type of plaque assay is in standard use for bacterial and animal viruses. Plant viruses can be assayed similarly by counting local lesions on plant leaves inoculated with viruses. Analysis of viral mutants, which are commonly isolated by plaque assays, has contributed extensively to current understanding of molecular cellular processes. The plaque assay also is critical in isolating *l* bacteriophage clones carrying segments of cellular DNA.

VIRAL GROWTH CYCLES

The surface of viruses includes many copies of one type of protein that binds, or adsorbs, specifically to multiple copies of a receptor protein on a host cell. This interaction determines the host range of a virus and begins the infection process. Then, in one of various ways, the viral DNA or RNA crosses the plasma membrane into the cytoplasm.

The entering genetic material may still be accompanied

by inner viral proteins, although in the case of many bacteriophages, all capsid proteins remain outside an infected cell. The genome of most DNA-containing viruses that infect eukaryotic cells is transported (with some associated proteins) into the cell nucleus, where the cellular DNA is, of course, also found. Once inside the cell, the viral DNA interacts with the host's machinery for transcribing DNA into mRNA. The viral mRNA that is produced then is translated into viral proteins by host-cell ribosomes, tRNA, and translation factors.

Most viral protein products fall into one of three categories: special enzymes needed for viral replication; inhibitory factors that stop host-cell DNA, RNA, and protein synthesis; and structural proteins used in the construction of new virions. These last proteins generally are made in much larger amounts than the other two types. After the synthesis of hundreds to thousands of new virions has been completed, most infected bacterial cells and some infected plant and animal cells rupture, or lyse, releasing all the virions at once. In many plant and animal viral infections, however, no discrete lytic event occurs; rather, the dead host cell releases the virions as it gradually disintegrates.

These events — adsorption, penetration, replication, and release — describe the lytic cycle of viral replication. The outcome is the production of a new round of viral particles and death of the cell. Adsorption and release of enveloped animal viruses are somewhat more complicated processes. In this case, the virions "bud" from the host cell, thereby acquiring their outer phospholipid envelope, which contains mostly viral glycoproteins. We illustrate the lytic cycle of enveloped viruses with the rabies virus, whose nucleocapsid consists of a single-stranded RNA genome surrounded by multiple copies of nucleocapsid protein. Within the nucleocapsid of rabies virions are viral enzymes for synthesizing viral mRNA and replicating the viral genome. The envelope around the nucleocapsid is a phospholipid bilayer containing multiple copies of a viral transmembrane glycoprotein.

This receptor-binding, or "attachment," protein has a large

external folded domain on the outside of the viral envelope, an *a*-helical transmembrane domain that spans the viral envelope, and a short internal domain. The internal domain interacts with the viral matrix protein, which functions as a bridge between the transmembrane glycoprotein and nucleocapsid protein. Assembly of progeny nucleocapsids, and release of progeny virions by budding from the host-cell plasma membrane. Budding virions are clearly visible in electron micrographs.

In some cases, after a bacteriophage DNA molecule enters a bacterial cell, it becomes integrated into the host-cell chromosome, where it remains quiescent and is replicated as part of the cell's DNA from one generation to the next. This association is called lysogeny, and the integrated phage DNA is referred to as a *prophage*. Under certain conditions, the prophage DNA is activated, leading to its excision from the host-cell chromosome and entrance into the lytic cycle. Bacterial viruses of this type are called *temperate phages*.

The genomes of a number of animal viruses also can integrate into the host-cell genome. Probably the most important are the retroviruses, described briefly later in this chapter.

A few phages and animal viruses can infect a cell and cause new virion production without killing the cell or becoming integrated. Four Types of Bacterial Viruses Are Widely Used in Biochemical and Genetic Research Bacterial viruses have played a crucial role in the development of molecular cell biology.

Thousands of different bacteriophages have been isolated; many of these are particularly well suited for studies of specific biochemical or genetic events. Here, we briefly describe four types of bacteriophages, all of which infect *E. coli*, that have been especially useful in molecular biology research.

DNA PHAGES OF THE T SERIES

The T phages of E. coli are large lytic phages that contain a single molecule of double-stranded DNA. This molecule is about 2×10^5 base pairs long in T2, T4, and T6 viruses and about 4 × 10^4 base pairs long in T1, T3, T5, and T7 viruses. T-phage virions consist of a helical protein "tail" attached to an icosahedral "head" filled with the viral DNA. After the tip of a T-phage tail adsorbs to receptors on the surface of an E. coli cell, the DNA in the head enters the cell through the tail.

The phage DNA then directs a programme of events that produces approximately 100 new phage particles in about 20 minutes, at which time the infected cell lyses and releases the new phages. The initial discovery of the role of messenger RNA in protein synthesis was based on studies of E. coli cells infected with bacteriophage T2. By 20 minutes after infection, infected cells synthesize T2 proteins only. The finding that the RNA synthesized at this time had the same base composition as T2 DNA (not E. coli DNA) implied that mRNA copies of T2 DNA were synthesized and used to direct cellular ribosomes to synthesize T2 proteins.

TEMPERATE PHAGES

Bacteriophage l, which infects E. coli, typifies the temperate phages. This phage has one of the most studied genomes and is used extensively in DNA cloning. On entering an E. coli cell, the double-stranded l DNA assumes a circular form, which can enter either the lytic cycle (as T phages do) or the lysogenic cycle. In the latter case, proteins expressed from the viral DNA bind a specific sequence on the circular viral DNA to a similar specific sequence on the circular bacterial DNA.

The viral proteins then break both circular molecules of DNA and rejoin the broken ends, so that the viral DNA becomes inserted into the host DNA. The carefully controlled action of viral genes maintains l DNA as part of the host chromosome by repressing the lytic functions of the phage. Under appropriate stimulation, the l prophage is activated and undergoes lytic replication.

SMALL DNA PHAGES

The genome of some bacteriophages encodes only 10 – 12 proteins, roughly 5 – 10 per cent of the number encoded by T phages. These small DNA phages are typified and the filamentous M13 phages. These were the first organisms in which the entire DNA sequence of a genome was determined, permitting extensive understanding of the viral life cycle. The viruses in this group are so simple that they do not encode most of the proteins required for replication of their DNA but depend on cellular proteins for this purpose. For this reason, they have been particularly useful in identifying and analyzing the cellular proteins involved in DNA replication.

RNA PHAGES

Some E. coli bacteriophages contain a genome composed of RNA instead of DNA. Because they are easy to grow in large amounts and because their RNA genomes also serve as their mRNA, these phages are a ready source of a pure species of mRNA. In one of the earliest demonstrations that cell-free protein synthesis can be mediated by mRNA, RNA from these phages was shown to direct the synthesis of viral coat protein when added to an extract of E. coli cells containing all the other components needed for protein synthesis. Also, the first long mRNA molecule to be sequenced was the genome of an RNA phage. These viruses, among the smallest known, encode only four proteins: an RNA polymerase for replication of the viral RNA, two capsid proteins, and an enzyme that dissolves the bacterial cell wall and allows release of the intracellular virus particles into the medium.

Animal Viruses are Classified by Genome Type

Animal viruses come in a variety of shapes, sizes, and genetic strategies. In this book, we are concerned with viruses that exhibit at least one of two features: they utilize important cellular pathways to form their molecules, thereby closely mimicking a normal cellular function, or they can integrate their genomes into those of normal cells.

The names of many viruses are based on the names of the diseases they cause or of the animals or plants they infect. Common examples include poliovirus, which causes poliomyelitis; tobacco mosaic virus, which causes a mottling disease of tobacco leaves; and human immunodeficiency virus (HIV), which causes acquired immunodeficiency syndrome (AIDS).

However, many different kinds of viruses often produce the same symptoms or the same apparent disease states; for example, several dozen different viruses can cause the red eyes, runny nose, and sneezing referred to as the common cold. Clearly, any attempt to classify viruses on the basis of the symptoms they produce or their hosts obscures many important differences in their structures and life cycles.

What *are* central to the life cycle of a virus are the types of nucleic acids formed during its replication and the pathway by which mRNA is produced.

The relation between the viral mRNA and the nucleic acid of the infectious particle is the basis of a simple means of classifying viruses.

In this system, a viral mRNA is designated as a *plus strand* and its complementary sequence, which cannot function as an mRNA, is a *minus strand.* A strand of DNA complementary to a viral mRNA is also a minus strand. Production of a plus strand of mRNA requires that a minus strand of RNA or DNA be used as a template.

Using this system, six classes of animal viruses are recognized. Bacteriophages and plant viruses also can be classified in this way, but the system has been used most widely in animal virology because representatives of all six classes have been identified.

DNA VIRUSES (CLASSES I AND II)

The genomes of both class I and class II viruses consist of DNA. Various types of DNA viruses are commonly used in studies on DNA replication, genome structure, mRNA production, and oncogenic cell transformation. *Class I viruses* contain a single molecule of double-stranded DNA (dsDNA). In the case of the most common type of class I animal virus, viral DNA enters the cell nucleus, where cellular enzymes transcribe the DNA and process the resulting RNA into viral mRNA.

Examples of these viruses include the following:

- *Adenoviruses,* which cause infections in the upper respiratory tract and gastrointestinal tract in many animals
- *SV40* (simian virus 40), a monkey virus that was accidentally discovered in kidney cell cultures from wild monkeys used in the production of poliovirus vaccines
- *Herpesviruses,* which cause various inflammatory skin diseases (e.g., chickenpox) and latent infections that recur after long intervals (e.g., cold sores and shingles)
- *Human papillomaviruses* (HPVs), which cause warts and other insignificant skin lesions and occasionally cause malignant transformation of cervical cells

Some types of HPV are passed through sexual contact. In some infected women, the HPV genome integrates into the chromosome of a cervical epithelial cell.

This rare integration event initiates an intensively studied process that can lead to development of cervical carcinoma, one of the most common types of human cancers. Routine Pap smears performed for early detection of cervical carcinoma are done to identify cells in the early stages of the transformation process initiated by HPV integration.

The second type of class I virus, collectively referred to as *poxviruses,* replicates in the host-cell cytoplasm. Typical of class Ib viruses are variola, which causes smallpox, and vaccinia, an attenuated (weakened) poxvirus used in vaccinations to induce immunity to smallpox. These very large, brick-shaped viruses ($0.1 \times 0.1 \times 0.2\ \mu m$) carry their own enzymes for synthesizing viral mRNA and DNA in the cytoplasm. *Class II viruses,* called *parvoviruses* (from Latin *parvo,* "poor"), are simple viruses that contain one molecule of single-stranded DNA (ssDNA). Some parvoviruses encapsidate (enclose) both plus and minus strands of DNA, but in separate virions; others encapsidate only the minus strand. In both cases, the ssDNA is copied inside the cell into dsDNA, which is then itself copied into mRNA.

RNA VIRUSES (CLASSES III – VI)

All the animal viruses belonging to classes III – VI have RNA genomes. A wide range of animals, from insects to human beings, are infected by viruses in each of these classes. These viruses have been particularly useful in studies on mRNA synthesis and translation (class III); glycoprotein synthesis, membrane formation, and intracellular transport (classes IV and V); and cell transformation and oncogenes (class VI).

Class III viruses contain double-stranded genomic RNA (dsRNA). The minus RNA strand acts as a template for the synthesis of plus strands of mRNA. The virions of all class III viruses known to date have genomes containing 10 – 12 separate double-stranded RNA molecules, each of which encodes one or two polypeptides.

Consequently, these viruses are said to have "segmented" genomes. In these viruses, the virion itself contains a complete set of enzymes that can utilize the minus strand of the genomic RNA as a template for synthesis of mRNA in the test tube as well as in the cell cytoplasm after infection. A number of important studies have used class III viruses as a source of pure mRNA.

Class IV viruses contain a single plus strand of genomic RNA, which is identical with the viral mRNA. Since the genomic RNA encodes proteins, it is infectious by itself. During replication of class IV viruses, the genomic RNA is copied into a minus strand, which then acts as a template for synthesis of more plus strands, or mRNA.

Two types of class IV viruses are known. In class IVa viruses, typified by poliovirus, viral proteins are first synthesized, from a single mRNA species, as a long polypeptide chain, or *polyprotein,* which is then cleaved to yield the various functional proteins.

Class IVb viruses synthesize at least two species of mRNA in a host cell. One of these mRNAs is the same length as the virion's genomic RNA; the other corresponds to the 32 third of the genomic RNA. Both mRNAs are translated into polyproteins.

Included in class IVb are a large number of rare insect-borne viruses including Sindbis virus and those causing yellow fever and viral encephalitis in human beings. These viruses once were called *arboviruses* (arthropod-borne viruses), but now are called *togaviruses* (from Latin *toga,* cover) because the virions are surrounded by a lipid envelope.

Class V viruses contain a single negative strand of genomic RNA, whose sequence is complementary to that of the viral mRNA. The genomic RNA in the virion acts as a template for synthesis of mRNA but does not itself encode proteins. Two types of class V viruses can be distinguished.

The genome in class Va viruses, which include the viruses causing measles and mumps, is a single molecule of RNA. A virus-specific RNA polymerase present in the virion catalyzes synthesis of several mRNAs, each encoding a single protein, from the genomic template strand.

Class Vb viruses, typified by influenza virus, have segmented genomes; each segment acts as a template for the synthesis of a different mRNA species. In most cases, each mRNA produced by a class Vb virus encodes a single protein; however, some mRNAs can be read in two different frames to yield two distinct proteins.

As with class Va viruses, a class Vb virion contains a virus-specific polymerase that catalyzes synthesis of the viral mRNA. Thus the genomic RNA (a minus strand) in both types of class V viruses is not infectious in the absence of the virus-specific polymerase.

The influenza RNA polymerase initiates synthesis of each mRNA by a unique mechanism. In the host-cell nucleus, the polymerase cuts off 12 – 15 nucleotides from the 52 end of a cellular mRNA or mRNA precursor; this oligonucleotide acts as a "primer" that is elongated by the polymerase to form viral (+) mRNAs, using the genomic (") RNA as a template.

Class VI viruses are enveloped viruses whose genome consists of two identical plus strands of RNA. These viruses are also known as retroviruses because their RNA genome directs the formation of a DNA molecule.

The DNA molecule ultimately acts as the template for

synthesis of viral mRNA. Initially, a viral enzyme called reverse transcriptase copies the viral RNA genome into a single minus strand of DNA; the same enzyme then catalyzes synthesis of a complementary plus strand. The resulting dsDNA is integrated into the chromosomal DNA of the infected cell.

Finally, the integrated proviral DNA is transcribed by the cell's own machinery into (+) RNA, which either is translated into viral proteins or is packaged within virion coat proteins to form progeny virions, which are released by budding from the host-cell membrane.

Because most retroviruses do not kill their host cells, infected cells can replicate, producing daughter cells with integrated proviral DNA. These daughter cells continue to transcribe the proviral DNA and bud progeny virions. Some retroviruses contain cancer-causing genes (called oncogenes). Cells infected by such retroviruses are oncogenically transformed into tumor cells. Studies of oncogenic retroviruses (mostly viruses of birds and mice) have revealed a great deal about the processes that lead to oncogenic transformation. Among the known human retroviruses are human T-cell lymphotrophic virus (HTLV), which causes a form of leukemia, and human immunodeficiency virus (HIV), which causes acquired immune deficiency syndrome (AIDS).

Both of these viruses can infect only specific cell types, primarily certain cells of the immune system and, in the case of HIV, some central nervous system neurons and glial cells. Only these cells have cell-surface receptors that interact with viral proteins, accounting for the host-cell specificity of these viruses.

VIRAL VECTORS

Knowledge about mechanisms of viral replication has allowed virologists to modify viruses for various purposes. For instance, the ability of virions to introduce their contents into the cytoplasm and nuclei of infected cells has been adapted for use in DNA cloning and offers possibilities in the treatment of certain diseases.

The introduction of new genes into cells by packaging them into virion particles is called *viral gene transduction,* and the virions used for this purpose are called *viral vectors.*

By use of recombinant DNA techniques it is a relatively straight forward process to construct human adenovirus *recombinants* in which potentially therapeutic genes replace the viral genes required for the lytic cycle of infection. Because adenovirus has a very broad host range for different types of human cells, these vectors can introduce the engineered gene into the cells of tissues where they are applied.

If the transduced gene encodes the normal form of a protein that is missing or defective in a particular disease, then such *gene therapy* may successfully treat the disease.

One type of adenovirus, for example, efficiently infects cells lining the air passages in the lungs, causing a type of common cold. Researchers have replaced some of the disease-causing genes in this adenovirus with the *CFTR* gene, which is defective in individuals with cystic fibrosis.

This recombinant adenovirus currently is being used to introduce a normal *CFTR* gene into the airway-lining cells of cystic fibrosis patients.

Unfortunately, with most of the adenovirus vectors currently available, the transduced gene usually is expressed only for a limited period of 2 to 3 weeks. This significantly limits their usefulness in gene therapy.

Viral vectors have also been developed from viruses that integrate their genomes into host-cell chromosomes. Such vectors have the advantage that progeny of the initially infected cell also contain and express the transduced gene because it is replicated and segregated to daughter cells along with the rest of the chromosome into which it is integrated. Retroviral vectors, which can efficiently integrate transduced genes at approximately random positions in host-cell chromosomes are now widely used experimentally to generate cultured cells expressing specific, desired proteins.

However, technical limitations in producing the large numbers of retroviral vectors required to infect a significant fraction of cells in the tissues of a human or vertebrate currently limit their use as gene therapy vectors. Another concern with retroviral vectors is that their random integration might disturb the normal expression of cellular genes encoding proteins regulating cellular replication.

This type of cellular gene deregulation occurs naturally following infection with certain retroviruses, such as avian leukosis virus and murine leukemia viruses, leading to development of leukemia in birds and mice, respectively.

Adeno-associated virus (AAV) is a "satellite" parvovirus that replicates only in cells that are co-infected with adenovirus or herpes simplex virus.

When AAV infects human cells in the absence of these "helper" viruses, its ssDNA genome is copied into dsDNA by host-cell DNA polymerase and then is integrated into a single region on chromosome, where it does not have any known deleterious effects. Research is under way to adapt the AAV integration mechanism that operates in the absence of helper virus to the development of a safe and effective integrating viral vector.

SUBCELLULAR ORGANIZATION OF EUKARYOTIC CELLS

Recognition that modern-day organisms fall into three distinct evolutionary lineages came from comparisons of the nucleotide sequences of genes common to all organisms and of ribosomal RNAs, particularly the RNA species found in the small ribosomal subunit. Prokaryotes, which lack a defined nucleus and have a simple subcellular organization, form two of the lineages — the bacteria and archaea.

The third lineage is the eukaryotes (eukarya), whose cells have a membrane-limited nucleus containing most of the cellular DNA, numerous specialized organelles, and a complex cytoskeleton. Despite the differences in cellular organization between prokaryotes and eukaryotes, all cells share certain structural features and carry out DNA replication, protein synthesis, production of ATP, and many other complicated metabolic events in basically the same way.

At the molecular level there is a close relationship between protein structure and function. Likewise, at the cellular level, function and structure are intimately related. Thus this chapter describes the basic structural components of cells and the general functions of each, focusing on eukaryotic cells.

We begin by describing the capabilities and applications of various techniques of microscopy used to visualize cells and subcellular structures. We then consider the basic methods for purifying biological membranes and subcellular structures. The purified preparations produced by cell fractionation permit detailed studies on the structure and function of biomembranes and other subcellular structures. Following this overview of experimental techniques, we continue the discussion of biomembranes begun in earlier chapters.

The surface membrane found on all cells and the membranes that line eukaryotic organelles all have the same basic architecture — a phospholipid bilayer; the unique function of each type of membrane is determined primarily by the proteins it contains.

The final section briefly describes the structure and function of the main internal organelles and fibers of eukaryotic cells, as well as the extracellular substances that surround cells and give them shape and strength. This review sets the stage for later discussions of cell function at the molecular level.

SAMPLES FOR LIGHT MICROSCOPY

Specimens for light microscopy are commonly fixed with a solution containing alcohol or formaldehyde, compounds that denature most proteins and nucleic acids. Formaldehyde also cross-links amino groups on adjacent molecules; these covalent bonds stabilize protein-protein and protein – nucleic acid interactions and render the molecules insoluble and stable for subsequent procedures.

Usually the sample is then embedded in paraffin or plastic and cut into thin sections of one or a few micrometers thick. Alternatively, the sample can be frozen without prior fixation and then sectioned; this avoids the denaturation of enzymes by fixatives such as formaldehyde.

Since the resolution of the light microscope is ≈0.2 μm and mitochondria and chloroplasts are ≈1 μm long (about the size of bacteria), theoretically one should be able to see these organelles.

However, most cellular constituents are not coloured and absorb about the same degree of visible light, so that they are hard to distinguish under a light microscope unless the specimen is stained. Thus the final step in preparing a specimen for light microscopy is to stain it, in order to visualize the main structural features of the cell or tissue. Many chemical stains bind to molecules that have specific features. For example, *hematoxylin* binds to basic amino acids (lysine and arginine) on many different kinds of proteins, whereas *eosin* binds to acidic molecules (such as DNA, and aspartate and glutamate side chains).

Because of their different binding properties, these dyes stain various cell types sufficiently differently that they are distinguishable visually. Two other common dyes are *benzidine*, which binds to heme-containing proteins and nucleic acids, and *fuchsin*, which binds to DNA and is used in Fuelgen staining. If an enzyme catalyzes a reaction that produces a coloured or otherwise visible precipitate from a colourless precursor, the enzyme may be detected in cell sections by their coloured reaction products. This technique is called *cytochemical staining*.

Fluorescence Microscopy Can Localize Specific Molecules in Cells

Perhaps the most versatile and powerful technique for localizing proteins within a cell by light microscopy is fluorescent staining of cells and observation in the *fluorescence microscope*. A chemical is said to be *fluorescent* if it absorbs light at one wavelength (the *excitation wavelength*) and emits light (fluoresces) at a specific and longer wavelength.

Most fluorescent dyes emit visible light, but some (such as Cy5 and Cy7) emit infrared light. In modern fluorescence microscopes, only fluorescent light emitted by the sample is used to form an image; light of the exciting wavelength induces the fluorescence but is then not allowed to pass the filters placed between the objective lens and the eye or camera.

REVEALING SPECIFIC PROTEINS IN FIXED CELLS

Four very useful dyes for fluorescent staining are rhodamine and Texas red, which emit red light; Cy3, which emits orange light; and fluorescein, which emits green light. These dyes have a low, nonspecific affinity for biological molecules, but they can be chemically coupled to purified antibodies specific for almost any desired macromolecule.

When a fluorescent dye – antibody complex is added to a permeabilized cell or tissue section, the complex will bind to the corresponding antigens, which then light up when illuminated by the exciting wavelength, a technique called *immunofluorescence microscopy*. By staining a specimen with two or three dyes that fluoresce at different wavelengths, multiple proteins can be localized within a cell.

REVEALING SPECIFIC PROTEINS IN LIVING CELLS

Fluorescence microscopy can also be applied to live cells. For example, purified actin may be chemically linked to a fluorescent dye. Careful biochemical studies have established that this "tagged" molecule is indistinguishable in function from its normal counterpart.

If the tagged protein is *microinjected* into a cultured cell, the endogenous cellular and injected tagged actin monomers copolymerize into normal long actin fibers. This technique can also be used to study individual microtubules within a cell.

Another technique for detecting specific proteins within living cells takes advantage of *green fluorescent protein* (GFP), a naturally fluorescent protein found in the jellyfish *Aequorea victoria*. The bioluminescence of this organism, which radiates a green fluorescence, is due to GFP.

This 238-aa protein contains serine, tyrosine, and glycine residues whose side chains have spontaneously reacted with one another to form a fluorescent chromophore. By recombinant DNA techniques the GFP gene can be introduced into living cultured cells or into specific cells of an entire animal. Because the introduced gene will express GFP, the cells will emit a green fluorescence when irradiated; this GFP fluorescence can be used to localize the cells within a tissue.

Alternatively, the gene for GFP can be fused to the gene for another protein of interest, producing a recombinant DNA encoding one long chimeric protein that contains the entirety of both proteins. Cells in which this recombinant DNA has been introduced will synthesize this chimeric protein, whose green fluorescence will reveal the subcellular localization of the protein.

INTRACELLULAR CONCENTRATION OF Ca^{2+} AND H^+ IONS

Changes in the cytosolic concentration of Ca^{2+} ions or pH frequently signal changes in cellular metabolism. The Ca^{2+} concentration in the cytosol of resting cells, for instance, is about 10^{-7} M. Many hormones or other stimuli cause a rise in cytosolic Ca^{2+} to 10^{-6} M; this, in turn, causes changes in cellular metabolism, such as contraction of muscle.

The fluorescent properties of certain dyes, such as *fura-2*, facilitate measurement of the concentration of free Ca^{2+} in the cytosol. This dye contains five carboxylate groups that form ester linkages with ethanol. The resulting fura-2 ester is lipophilic and can diffuse from the medium across the plasma membrane into cells.

Within the cytosol, esterases hydrolyze fura-2 ester yielding fura-2, whose free carboxylate groups render the molecule nonlipophilic, so it cannot cross cellular membranes and remains in the cytosol.

Each fura-2 molecule can bind a single Ca^{2+} ion but no other cellular cation, and the amount of fura-2 bound to Ca^{2+} is proportional, over a certain range, to the Ca^{2+} concentration. The fluorescence of fura-2 at one particular wavelength is enhanced when Ca^{2+} is bound, and the fluorescence is proportional to the Ca^{2+} concentration.

At another wavelength the fluorescence of fura-2 is the same whether or not Ca^{2+} is bound and provides a measure of the total amount of fura-2 in the segment of the cell. By examining cells continuously in the fluorescence microscope and measuring rapid changes in the ratio of fura-2 fluorescence at these two wavelengths, one can quantify rapid changes in the fraction of fura-2 that has a bound Ca^{2+} ion and thus in the concentration of cytosolic Ca^{2+}. The fluorescence of other dyes is sensitive to the H^+ concentration and can be used in a similar way to monitor the cytosolic pH of living cells.

CONFOCAL SCANNING AND DECONVOLUTION MICROSCOPY

Immunofluorescence microscopy has its limitations. The fixatives employed to preserve cell architecture often destroy the *antigenicity* of a protein, that is, its ability to bind to its specific antibody. Also, the method generally gives poor results with thin cell sections, because embedding media often fluoresce themselves, obscuring the specific signal from the antibody.

Moreover, in microscopy of whole cells, the fluorescent light comes from molecules above and below the plane of focus; thus the observer sees a superposition of fluorescent images from molecules at many depths in the cell, making it difficult to determine the actual three-dimensional molecular arrangement.

The *confocal scanning microscope* avoids the last problem by permitting the observer to visualize fluorescent molecules in a single plane of focus, thereby creating a vastly sharper cross-sectional image. At any instant during confocal imaging, only a single small part of a sample is illuminated with exciting light from a focused laser beam, which rapidly moves to different spots in the sample focal plane.

Images from these spots are recorded by a video camera and stored in a computer, and the composite image is displayed on a computer screen.

Deconvolution microscopy is similar to confocal microscopy in that a cross-sectional image is obtained, but the two techniques differ in the details of how this image is generated. In both cases, the objective lens collects light that originates from above and below the focal plane as well as that which originates from within the focal plane.

Confocal microscopes use a pinhole to exclude the out-of-focus light. In contrast, deconvolution microscopes collect all the light from several focal planes, and then mathematically reassign the out-of-focus light to its correct focal plane with the aid of a high-speed computer, a mathematical operation called *deconvolution*.

To understand how a deconvolution microscope works, consider an infinitely small fluorescent source of light, which can be approximated by a fluorescent bead smaller than the resolution of the light microscope (i.e., <0.2 μm in diameter). The emitted light radiates in all directions, and when the source is in the focal plane of the objective, it appearsas a bright point of light.

When the point source is outside the focal plane of the objective, some of the light is still collected by the objective lens, and the point source appears as a halo. As the focal plane is moved farther away from the plane containing the point source, the halo becomes larger and more diffuse.

Knowing exactly how the light emitted by an infinitely small fluorescent source is collected and distorted by the optics of the sample and microscope, it is possible to reconstruct an individual cross-sectional image (containing only light that originated in the focal plane of interest) from a set of images taken as the objective focal plane is moved through the plane of interest.

Cross-sectional images obtained with a deconvolution microscope may have even greater detail than those obtained with a confocal microscope. Additionally, the fluorescent labeling of the sample does not need to be as intense for deconvolution microscopy as it does for confocal microscopy, since all the light produced by a fluorescent sample is collected and analyzed by the microscope.

Three-dimensional images can be obtained by a refinement known as *optical sectioning*. In this method, a computer records individual fluorescent images of planes at different depths of the sample — in effect, serial sections —

and combines the stack of images into one three-dimensional image.

MICROSCOPY VISUALIZE UNSTAINED LIVING CELLS

Detailed views of transparent, live, unstained cells and tissues are obtainable with *phase-contrast microscopy* and *Nomarski interference microscopy*. Both techniques take advantage of the phenomena of refraction and diffraction of light waves. As a result, small differences in refractive index and thickness between parts of the specimen (say, between the nucleus and cytosol) or between the specimen and the surrounding medium can be converted into differences of light and dark in the final image. The phase-contrast microscope generates an image in which the degree of darkness or brightness of a region of the sample depends on the refractive index of that region. The improved definition of subcellular structures in live, unstained cells obtained by phase-contrast microscopy compared with standard bright-field microscopy.

Nomarski, or differential, interference microscopy generates an image that looks as if the specimen is casting a shadow to one side: the "shadow" primarily represents a difference in refractive index and thickness of a specimen rather than its topography. In this technique, a prism splits an incident beam of plane-polarized light so that one part of the beam passes through one region of a specimen and the other part passes through a closely adjacent region; a second prism then reassembles the two beams.

Minute differences in thickness or in the refractive index between adjacent parts of a sample are converted into a bright image (if the two beams are in phase when they recombine) or a dark one (if they are out of phase).

Phase-contrast microscopy is especially useful in examining the structure and movement of larger organelles, such as the nucleus and mitochondria, in live cultured cells. The greatest disadvantage of this technique is that it is suitable for observing only single cells or thin cell layers. Nomarski interference microscopy, in contrast, defines only the outlines of large organelles, such as the nucleus and vacuole. However, thick objects, such as the nuclei in a worm, can be observed by combining this technique with optical sectioning.

Both phase-contrast and Nomarski interference microscopy can be used in *time-lapse microscopy,* in which the same cell is photographed at regular intervals over periods of several hours. This procedure allows the observer to study cell movement, provided the microscope's stage can control the temperature of the specimen and the gas environment.

LIMIT OF TRANSMISSION ELECTRON MICROSCOPY

The fundamental principles of electron microscopy are similar to those of light microscopy; the major difference is that electromagnetic lenses, not optical lenses, focus a high velocity electron beam instead of visible light. Because electrons are absorbed by atoms in air, the entire tube between the electron source and the viewing screen is maintained under an ultrahigh vacuum.

The *transmission electron microscope* (TEM) directs a beam of electrons through a specimen. Electrons are emitted by a tungsten cathode when it is electrically heated. The electric potential of the cathode is kept at 50,000 – 100,000 volts; that of the anode, near the top of the tube, is zero. This drop in voltage causes the electrons to accelerate as they move toward the anode.

A condenser lens focuses the electron beam onto the sample; objective and projector lenses focus the electrons that pass through the specimen and project them onto a viewing screen or a piece of photographic film.

In typical electron microscopes, electrons have the properties of a wave with a wavelength of only 0.005 nm. Recall that the minimum distance D at which two objects can be distinguished is proportional to the wavelength λ of the light that illuminates the objects.

Thus the limit of resolution for the electron microscope is theoretically 0.005 nm (less than the diameter of a single atom), or 40,000 times better than the resolution of the light microscope and 2 million times better than that of the unaided human eye.

However, the effective resolution of the electron microscope in the study of biological systems is considerably less than this ideal. Under optimal conditions, a resolution of 0.10 nm can be obtained with transmission electron microscopes, about 2000 times better than the best resolution of light microscopes.

SAMPLES FOR TRANSMISSION ELECTRON MICROSCOPY

Like the light microscope, the transmission electron microscope is used to view thin sections of a specimen, but the fixed sections must be much thinner for electron microscopy (only 50 – 100 nm, about 0.2 per cent of the thickness of a single cell). Clearly, only a small portion of a cell can be observed in any one section. Generation of the image depends on differential scattering of the incident electrons by molecules in the preparation. Without staining, the beam of electrons passes through a cell or tissue sample uniformly, so the entire sample appears uniformly bright with little differentiation of components.

Staining techniques are therefore used to reveal the location and distribution of specific materials. Heavy metals, such as gold or osmium, appear dark on a micrograph because they scatter (diffract) most of the incident electrons; scattered electrons are not focused by the electromagnetic lenses and do not form the image.

Osmium tetrox-ide preferentially stains certain cellular components, such as membranes, which appear black in micrographs. Specific proteins can be detected in thin sections by use of electron-dense gold particles coated with protein A, a bacterial protein that binds antibody molecules nonspecifically.

Electron microscopy also is used to obtain information about the shapes of purified viruses, fibers, enzymes, and other subcellular particles. In one technique, called *metal shadowing*, a thin layer of evaporated metal, such as platinum, is laid at an angle on a biological sample. An acid bath dissolves the biological material, leaving a metal replica of its surface, which can then be examined in the transmission electron microscope.

Variations in the angle and thickness of the deposited metal allow an image to be formed because some incident electrons will be scattered in various directions rather than pass through the preparation.

If the metal is deposited mainly on one side of the sample, for instance, the image seems to have "shadows," where the metal appears dark and the shadows appear light.

CRYOELECTRON MICROSCOPY

Standard electron microscopy cannot be used to study live cells because they are generally too vulnerable to the required conditions and preparatory techniques. In particular, the absence of water causes macromolecules to become denatured and nonfunctional.

However, the technique of *cryoelectron microscopy* allows examination of hydrated, unfixed, and unstained biological specimens directly in the transmission electron microscope.

In this technique, an aqueous suspension of a sample is applied in an extremely thin film to a grid. After it has been frozen in liquid nitrogen and maintained in this state by means of a special mount, it is observed in the electron microscope.

The very low temperature keeps the water from evaporating, even in a vacuum, and the sample can be observed in detail in its native, hydrated state without shadowing or fixing it. By computer-based averaging of images of hundreds of particles, a three-dimensional model almost to atomic resolution can be generated.

SURFACES OF CELLS AND PARTICLES

The *scanning electron microscope* allows the investigator to view the surfaces of unsectioned specimens. These cannot be visualized with transmission equipment because the electrons pass through the entire specimen. The sample is fixed, dried, and coated with a thin layer of a heavy metal, such as platinum, by evaporation in a vacuum; in this case, the sample is rotated so that the platinum is deposited uniformly on the surface.

An intense electron beam inside the microscope scans rapidly over the sample. Molecules in the specimen are excited and release secondary electrons that are focused onto a scintillation detector; the resulting signal is displayed on a cathode-ray tube.

Because the number of secondary electrons produced by any one point on the sample depends on the angle of the electron beam in relation to the surface, the scanning electron micrograph has a threedimensional appearance. The resolving power of scanning electron microscopes, which is limited by the thickness of the metal coating, is only about 10 nm, much less than that of transmission instruments.

PURIFICATION OF CELLS AND THEIR PARTS

Most animal and plant tissues contain a mixture of cell types. However, an investigator often wishes to study a pure population of one type of cell. In some cases, cells differ in some physical property that allows different cell types to be separated.

White blood cells (leukocytes) and red blood cells (erythrocytes), for instance, have very different densities because erythrocytes have no nucleus; thus these cells can be separated on the basis of density. Since most cell types cannot be differentiated so easily, other cell-separation techniques have had to be developed. Similarly, it is essential to isolate quantities of each of the major subcellular organelles to study their structures and metabolic functions in detail.

FLOW CYTOMETRY SEPARATES DIFFERENT CELL TYPES

A *flow cytometer* can identify different cells by measuring the light they scatter, or the fluorescence they emit, as they flow through a laser beam; thus it can sort out cells of a particular type from a mixture. Indeed, a *fluorescence-activated cell sorter (FACS)*, an instrument based on flow cytometry, can select one cell from thousands of other cells.

For example, if an antibody specific to a certain cell-surface molecule is linked to a fluorescent dye, any cell bearing this molecule will bind the antibody and will then be separated from other cells when it fluoresces in the FACS. Once sorted from the other cells, the selected cell can be grown in culture. This procedure is commonly used to purify the different types of white blood cells, each of which bears on its surface one or more distinctive proteins and thus will bind monoclonal antibodies specific for that protein. Such FACS separations are more difficult to conduct on cultured cells or cells from animal tissues, which interact with adjacent cells and are surrounded by an extracellular matrix.

Samples must be treated with proteases to degrade the extracellular-matrix proteins and cell-surface proteins that attach cells in tissues to one another; these proteases usually also degrade the distinctive cell-surface "marker" proteins that distinguish one cell type from another.

Other uses of flow cytometry include the measurement of a cell's DNA and RNA content and the determination of its general shape and size. The FACS can make simultaneous measurements of the size of a cell (from the amount of scattered light) and the amount of DNA it contains (from the amount of fluorescence from a DNA-binding dye).

DISRUPTION OF CELLS RELEASES THEIR ORGANELLES

The initial step in purifying subcellular structures is to rupture the plasma membrane and the cell wall, if present. First, the cells are suspended in a solution of appropriate pH and salt content, usually isotonic sucrose (0.25 M) or a combination of salts similar in composition to those in the cell's interior.

Many cells can then be broken by stirring the cell suspension in a high-speed blender or by exposing it to highfrequency sound *(sonication)*. Plasma membranes can also be sheared by special pressurized tissue homogenizers in which the cells are forced through a very narrow space between the plunger and the vessel wall. Generally, the cell solution is kept at 0 °C to best preserve enzymes and other constituents after their release from the stabilizing forces of the cell. Because the plasma membrane is highly permeable to water but poorly permeable to the salts and other small molecules (solutes) within cells, *osmotic flow* can be enlisted to help rupture cells.

Recall that water flows across a semipermeable membrane, such as the plasma membrane, from a solution of high water (low solute) concentration to one of low water (high solute) concentration until the water concentration on both sides is equal. Consequently, when cells are placed in a hypotonic solution (i.e., one with a lower salt concentration than that of the cell interior), water flows into the cells. This osmotic flow causes the cells to swell and then more easily rupture. Conversely, in a hypertonic solution (i.e., one with a higher salt concentration than that of the cell interior), water flows out of cells, causing them to shrink.

When cells are placed in an isotonic solution (i.e., one with a salt concentration equal to that of the cell interior), there is no net movement of water in or out of cells. For this reason, an isotonic solution is best for preserving normal cell structure.

Disrupting the cell produces a mix of suspended cellular components, the *homogenate,* from which the desired organelles can be retrieved. Because rat liver contains an abundance of a single cell type, this tissue has been used in many classic studies of cell organelles. However, the same isolation principles apply to virtually all cells and tissues, and modifications of these cell-fractionation techniques can be used to separate and purify any desired components.

ORGANELLES CAN BE SEPARATED

We discussed the principles of centrifugation and the uses of centrifugation techniques for separating proteins and nucleic acids. Similar approaches are used for separating and purifying the various organelles, which differ in both size and density.

Most fractionation procedures begin with *differential centrifugation* at increasingly higher speeds also called *differential-velocity centrifugation.* The different sedimentation rates of various cellular components make it possible to separate them partially by centrifugation. Nuclei and viral particles can sometimes be purified completely by such a procedure. After centrifugation at each speed for an appropriate time, the supernatant is poured off and centrifuged at higher speed. Each pelleted fraction can be resuspended and further separated by equilibrium densitygradient centrifugation (discussed next).

Differential centrifugation does not yield totally pure organelle fractions. One method for further purifying fractions is *equilibrium density-gradient centrifugation,* which separates cellular components according to their density. The impure organelle fraction is layered on top of a solution that contains a gradient of a dense nonionic substance, such as sucrose or glycerol.

The tube is centrifuged at a high speed (about 40,000 rpm) for several hours, allowing each particle to migrate to an equilibrium position where the density of the surrounding liquid is equal to the density of the particle. In typical preparations from animal cells, the rough endoplasmic reticulum (density = 1.20 g/cm^3) separates well from the Golgi vesicles (density = 1.14 g/cm^3) and from the plasma membrane (density = 1.12 g/cm^3). (The higher density of the rough endoplasmic reticulum is due largely to the ribosomes bound to it.) This method also works well for resolving lysosomes, mitochondria, and peroxisomes in the initial mixed fraction obtained by differential centrifugation.

Since each organelle has unique morphological features, the purity of organelle preparations can be assessed by examination in an electron microscope. Alternatively, organelle-specific marker molecules can be quantified. For example, the protein cytochrome *c* is present only in mitochondria, so the presence of this protein in a fraction of lysosomes would indicate its contamination by mitochondria. Similarly, catalase is present only in peroxisomes; acid phosphatase, only in lysosomes; and ribosomes, only in the rough endoplasmic reticulum or the cytosol.

ORGANELLE-SPECIFIC ANTIBODIES ARE USEFUL IN PREPARING HIGHLY PURIFIED ORGANELLES

Cell fractions often contain more than one type of organelle even after differential and equilibrium density-gradient centrifugation. Such fractions can be further purified by immunological techniques, using monoclonal antibodies for various organelle-specific membrane proteins.

One example is the purification of a particular class of cellular vesicles whose outer surface is coated with the protein *clathrin*. An antibody to clathrin, bound to a bacterial carrier, can selectively bind these vesicles in a crude preparation of membranes, and the whole antibody complex can then be isolated by low-speed centrifugation. A recently developed technique uses tiny metallic beads coated with specific antibodies. Organelles that bind to the antibodies, and thus are linked to the metallic beads, are recovered from the preparation by adhesion to a small magnet on the side of the test tube.

All cells contain a dozen or more different types of small membrane-limited vesicles of about the same size (50–100 nm in diameter) and density. Because of their similar size and density, these vesicles are difficult to separate from one another by centrifugation techniques. Immunological techniques are particularly useful for purifying specific classes of such vesicles. Fat and muscle cells, for instance, contain a particular glucose transporter (GLUT4) that is localized to the membrane of a specific kind of vesicle. When insulin is added to the cells, these vesicles fuse with the cell-surface membrane, a process critical to maintaining the appropriate concentration of sugar in the blood. These vesicles can be purified using an antibody that binds to a segment of the GLUT4 protein that faces the cytosol.

UNIVERSALITY OF THE PHOSPHOLIPID BILAYER

A typical cell contains myriad types of membranes, each in turn bearing unique properties bestowed by its particular mix of lipids and proteins.

When samples of plasma, nuclear, and mitochondrial membranes are prepared, using the cell-fractionation techniques we have already described, these preparations are often contaminated with the membranes of many other organelles. (The plasma membrane of human erythrocytes, however, can be isolated in near purity because these cells contain no internal membranes.) And all cellular membranes, regardless of their source, possess enormously varied protein-to-lipid ratios.

The inner mitochondrial membrane, for example, is 76 per cent protein; the myelin membrane, only 18 per cent. The high phospholipid content of myelin allows it to electrically insulate the nerve cell from its environment.

Given the variable composition of cellular membranes, how certain are we that the phospholipid bilayer structure is common to all biomembranes? One piece of evidence is that many of the physical properties of pure phospholipid bilayers are similar to those of natural cellular membranes.

Another is that either a single species of phospholipid, or a mixture of phospholipids with a composition approximating that found in natural membranes, spontaneously forms either planar bilayers or liposomes when dispersed in aqueous solutions.

Perhaps the best evidence for the bilayer structure comes from low-angle x-ray diffraction analysis of the multimembrane myelin sheath, which is elaborated by Schwann cells and covers and insulates many mammalian nerve cells. The myelin sheath, which is a series of stacked membranes, is the major membrane component of such nerves and can be separated from other cellular membranes in a pure state, permitting direct physical and chemical analyses.

X-ray diffraction analysis of these stacked plasma membranes has revealed a regular variation in density that is consistent with a bilayer organization of each membrane unit. In this organization, protein is located mainly on either side of the membrane, which has hydrophilic external faces and a central region of almost pure low-density hydrocarbon. Although some polypeptide segments pass through the lipid bilayer, these make up less than 10 per cent of the inner mass of the membrane and are not detected in this type of analysis.

Electron microscopy of thin membrane sections stained with osmium tetroxide, which binds strongly to the polar head groups of phospholipids, provides the most direct evidence for the universality of the bilayer structure.

A cross section of all single membranes stained with osmium tetroxide looks like a railroad track: two thin dark lines (the stain – head group complexes) with a uniform light space of about 2 nm (the hydrophobic tails) between them. Although some osmium tetroxide may bind to double bonds in the hydrophobic fatty acyl chains, most of it binds to the polar head groups.

Chapter 8

Enzymes Activity

Enzymes are catalysts that optimize cell activity while minimizing the amount of energy needed to achieve a specific reaction. Enzymes are also energized protein molecules found in every living cell, and are necessary for life. There are over 2000 known enzymes, each of which is involved with one specific chemical reaction. They are any of various proteins, originating from living cells and capable of producing certain chemical changes in organic substances by catalytic action, such as digestion. These proteins, and their function (s), are determined by their shape.

In cells and organisms, most reactions are catalyzed by enzymes, which are regenerated during the course of a reaction. Biological catalysts are physiologically important because they speed up rates of reactions that would otherwise be too slow to support life. Our bodies naturally produce digestive and metabolic enzymes as they are needed. Specifically, the pancreas produces enzymes that break down foods into nutrients the body can use for energy and other bodily functions.

The names of enzymes often include the substrate or substance on which they act, joined with an-ase ending. For example, lactase acts upon lactose and maltase acts on maltose to produce glucose. Sometimes they are named for their reaction product, for example, sucrase is often called invertase, because invertase is the result of the reaction of sucrose. Enzymes can also bear a name the describes the reaction that is catalyzed. An example of this includes the use of the name oxidase, because oxidase is involved in an oxidation reaction.

TWO TYPES OF ENZYMES

- Simple enzymes are those composed completely of proteins
- Complex enzymes are those composed of protein and a small organic molecule (s) (also known as holoenzymes)

  Enzymes are Categorized by their Function (s):
- Metabolic enzymes catalyze and regulate every biochemical reaction that occurs in your body and are essential to cellular function and health.
- Digestive enzymes are secreted along the digestive tract, break down food into nutrients and waste, and allow the nutrients found in foods to be absorbed into the bloodstream. The pancreas produces most digestive enzymes, but the liver, gallbladder, small intestines, stomach and colon also play critical roles in the production of these enzymes.

  Examples of digestive enzymes: lipase, protease, amylase, ptyalin, pepsin and trypsin.
- Food enzymes enter the body through the consumption of raw foods or enzyme supplements. Cooking and processing destroys most enzymes in food.
- Plant Based enzymes are synthetically grown, help break down fat, protein and carbohydrates and function within a broad pH range. Plant enzymes, commonly bromelin (pineapple) and papain (papaya) play an important role in more complete digestion of all foods. They are activated at a temperature higher than normal body temperature, also making them a good anti-inflammatory.
- Proteoplytic enzymes are enzymes that digest proteins, including Trypsin, Chymotrypsin, Pancreatin, Bromelin and Papain. Some are produced by the pancreas and others are supplemental from animals or plants. The primary use of proteolytic enzymes in supplements is, are as digestive aids. They

are also used in drugs as anti-inflammatory agents and pain relievers.

Enzymes in Nature

Enzymes are one of the most interesting and important substances found in nature. Firts, it's important to realise that enzymes are not living things. They are inanimate-like minerals. But unlike minerals, they are made by living cells. If we were to look inside a cell we would see many different activities going on.

There would be some molecules joining together and others breaking apart. These activities keep the cell alive and enzymes make these activities possible. That's why every cell of every living creature on Earth produces enzymes. In a broad sense, there are two types of enzyme. Some that help join specific molecules together to form new molecules. Others that help break specific molecules apart into separate molecules.

Enzymes play many important roles ouside the cell as well. One of the best examples of this is the digestive system. For instance, it is enzymes in your digestive system that break food down in your digestive system break food down into small molecules that can be absorbed by the body. Some enzymes in your digestive system break down starch, some proteins and others break down fats.

Four Things to Remember About Enzymes

- *Enzymes are specific*: An enzyme that is able to break fat down would not be able to dissolve protein or starch. Enzymes perform only one specific job. That means an enzyme can do its job with very few side effects. It also explains why there are so many different types of enzyme. To date, 3,000 different types have been identified and there are many more waiting to be discovered.
- *Enzymes are catalysts*: While it is true that an enzyme can only perform one speific job, it is important to know that one enzyme can perform that same job over and over again, millions of times, without being

consumed in the process. And enzymes do their job best in the mild ph and temperature conditions found in nature.

- *Enzymes are efficient*: Not only do enzymes work hard, they also work with blinding speed. For instance, there is an enzyme in the liver that helps hydrogern peroxide break down into water and oxygen. What's amazing is that one enzyme can process 5 million hydrogen peroxide molecules in one minute.
- *Enzymes are natural*: Enzymes are proteins. Like all other proteins, enzymes are organic. Once they have done their job, enzymes break down swiftly and can be absorbed back into nature.

Enzymes in Industry

Many people believe enzyme technology is fairly new. However, that's not the case. Enzymes have been used by man from the dawn of civilization. As long as people have been eating bread and cheese, and drinking wine and beer-they have been using enzymes. That's because enzymes help to make these products.

Cheese

Take cheese for example. Most likely, cheese was discovered by ancient hunters who stored milk in the stomachs of slaughtered calves. When exposed to heat, the milk inside the container would turn into a solid-cheese. The reason for this is simple. Calves have an enzyme called chymosin in their stomachs. At ambient temperature chymosin will break down milk protein, causing milk to separate into curd and whey. If you strain away the watery whey you are left with a soft tangy curd or cheese. Chymosin derived from calf stomachs has been used in cheese making ever since.

Detergent

In more recent times, other commercial uses for enzymes have been found. Laundry detergents are a good example. As mentioned, there are certain enzymes which dissolve proteins and others which dissolve fat. Proteins and fat make up two of the major causes of stains on cloting. Grass, blood and egg are all protein stains. Lipstick, frying oil, butter, sauces, and tough stains on cuffs and collars are fat stains. to remove these stains without enzymes is difficult and requires a lot of washing at high temperatures with a lot of detergent.

Enzymes can remove these stains better, faster and in a way that is a lot kinder to the clothing and the environment. The enzymes in detergent dissolve stains in the same natural way they help digest food in your stomach-specifically and efficintly. Just a tiny amount of enzymes can dissolve stains that would require much larger amounts of detergent alone. Because of this they can reduce the amount of detergent actually found in laundry detergent. And since enzymes work best in mild conditions, washing machines using enzymatic detergants can be set at lower tempreatures, reducing electricity consumption by as much as a third. In the future, enzymes could replace more and more of the harsh chemicals found in detergents.

Many other Industries

These two examples represent just the tip of the iceberg. Enzymes are contributing to industry in many other areas. They are starting to replace petroleum-based solvents used to make vegetable seed oil; replacing harsh acids used in the production of glucose products such as corn syrup; taking over part of the role played by clorine in the paper industry; and replacing sulphides used in the tanning industry. Enzymes are also being used instead of pumice stones in the stonewashing of jeans.

Enzymes help Industry and Nature

Today enzymes are helping industry make the products used by society in a way that is less harmful to the environment. In the cases mentioned above, enzymatic processes are, in general, safer and more environment-friendly than the traditional processes they replace. That is why enzymes in particular and biotechnology in general hold such promise.

How Enzymes are Produced

Looking back to the cheese example above, chymosin is an enzyme used to turn milk into cheese. And chymosin is only found in the stomachs of young calves (as well as young sheep, goats, and a few other farm animals). Up until the 60s all the world's cheese was made from chymosin extracted from the stomachs of slaughtered calves. Then two things happened. The demand for cheese increased and the demand for alf meat decreased.

Soon there were not enough slaughtered calf stomachs to produce enough high-quality chymosin required by the expanding cheese industry. To solve the chymosin shortage, it would have been very wasteful and very expensive to raise and slaughter calves just to extract a relatively small amount of chymosin from their stomachs. The cheese industry and the enzyme researchers began to look for another type of organism that produced chymosin.

Microorganisms are Natural Enzyme Factories

The researchers wanted to find an organism that was easier and less expensive to raise than a calf. An organism that reproduced quickly and wouldn't require a lot of space and food. So they began looking among the smallest and simplest organisms they knew: micoorganisms. Microorganisms are very small living organisms. Some examples are bacteria, fungi and yeast. They live in soil or water in every corner of the earth. Because they are so small, they are obviously a lot less complex than a calf. Even so, a simple microorganism can produce many different types of enzyme. In fact, it doesn't take a complex organism to create a complex enzyme-as was proved by a simple fungus named Mucor. This fungus produced a chymosin-like enzyme that was almost the same as that produced by calves. But finding the right enzyme is only half the battle.

Microorganisms often aren't Suited for "life on the Farm"

Once identified a microorganism needs to be "farmed"

in a controlled setting. Most microorganisms can grow and multiply quite well in liquid. So large tanks are used called fermentation tanks to grow microorganisms. Ideally, a microorganism will grow fast and produce a lot of the desired enzyme at mild temperatures consuming inexpensive nutrients. But like most things in life, the ideal microorganism is hard to come by. As it is, most microorganisms found in the wild are not so well suited for domestication in fermentation tanks.

Some only produce tiny quantities of the desired enzyme, or they may produce undesirable by-products, or take a long time to grow, or require a growth environment that is very difficult and expensive to maintain. The result is that although we often find ourselves with a microorganism that can produce the enzyme needed the cost for cultivating it is prohibitive. That's where genetic engineering can help

Function and Structure

Enzymes are very efficient catalysts for biochemical reactions. They speed up reactions by providing an alternative reaction pathway of lower activation energy. Like all catalysts, enzymes take part in the reaction-that is how they provide an alternative reaction pathway.

But they do not undergo permanent changes and so remain unchanged at the end of the reaction. They can only alter the rate of reaction, not the position of the equilibrium. Most chemical catalysts catalyse a wide range of reactions. They are not usually very selective. In contrast enzymes are usually highly selective, catalysing specific reactions only.

This specificity is due to the shapes of the enzyme molecules. Many enzymes consist of a protein and a non-protein (called the cofactor). The proteins in enzymes are usually globular.

The intra-and intermolecular bonds that hold proteins in their secondary and tertiary structures are disrupted by changes in temperature and pH. This affects shapes and so the catalytic activity of an enzyme is pH and temperature sensitive.

Cofactors may be:

- Organic groups that are permanently bound to the enzyme (prosthetic groups)
- Cations-positively charged metal ions (activators), which temporarily bind to the active site of the enzyme, giving an intense positive charge to the enzyme's protein
- Organic molecules, usually vitamins or made from vitamins (coenzymes), which are not permanently bound to the enzyme molecule, but combine with the enzyme-substrate complex temporarily.

HOW ENZYMES WORK

For two molecules to react they must collide with one another. They must collide in the right direction (orientation) and with sufficient energy.

Sufficient energy means that between them they have enough energy to overcome the energy barrier to reaction. This is called the activation energy.

Enzymes have an active site. This is part of the molecule that has just the right shape and functional groups to bind to one of the reacting molecules. The reacting molecule that binds to the enzyme is called the substrate. An enzyme-catalysed reaction takes a different 'route'. The enzyme and substrate form a reaction intermediate. Its formation has a lower activation energy than the reaction between reactants without a catalyst.

A simplified picture:

Route Areactant 1 + reactant 2 → product

Route Breactant 1 + enzyme → intermediate

intermediate + reactant 2 → product + enzyme

So the enzyme is used to form a reaction intermediate, but when this reacts with another reactant the enzyme reforms.

LOCK AND KEY HYPOTHESIS

This is the simplest model to represent how an enzyme works. The substrate simply fits into the active site to form a reaction intermediate.

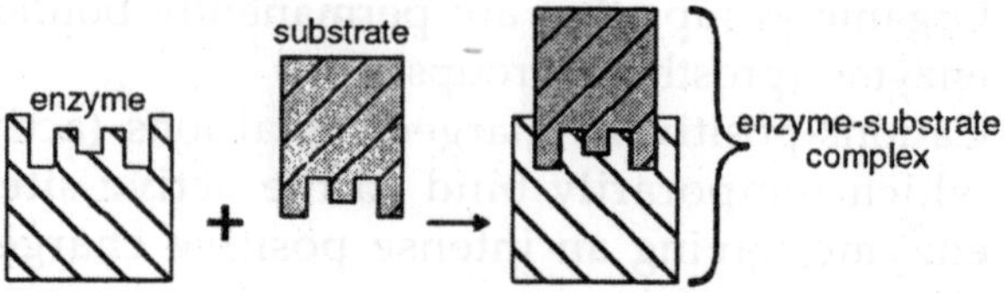

Induced Fit Hypothesis

In this model the enzyme molecule changes shape as the substrate molecules gets close. The change in shape is 'induced' by the approaching substrate molecule. This more sophisticated model relies on the fact that molecules are flexible because single covalent bonds are free to rotate.

Factors Affecting Catalytic Activity of Enzymes

Temperature

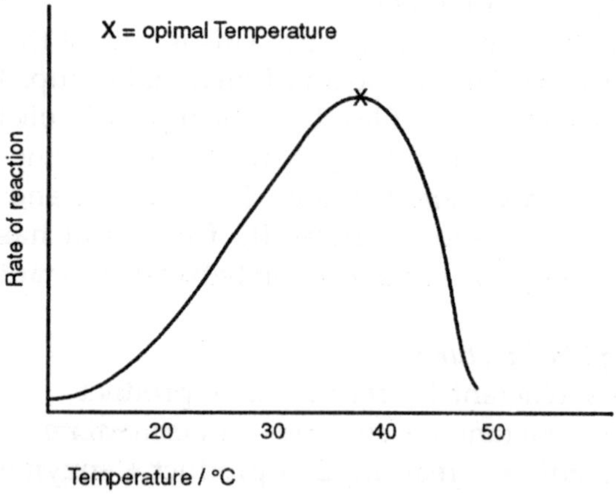

As the temperature rises, reacting molecules have more and more kinetic energy. This increases the chances of a successful collision and so the rate increases. There is a certain temperature at which an enzyme's catalytic activity is at its greatest.

This optimal temperature is usually around human body temperature (37.5 °C) for the enzymes in human cells.

pH

Below this temperature the enzyme structure begins to break down (denature) since at higher temperatures intra-and intermolecular bonds are broken as the enzyme molecules gain even more kinetic energy. Each enzyme works within quite a small pH range. There is a pH at which its activity is greatest (the optimal pH). This is because changes in pH can make and break intra-and intermolecular bonds, changing the shape of the enzyme and, therefore, its effectiveness.

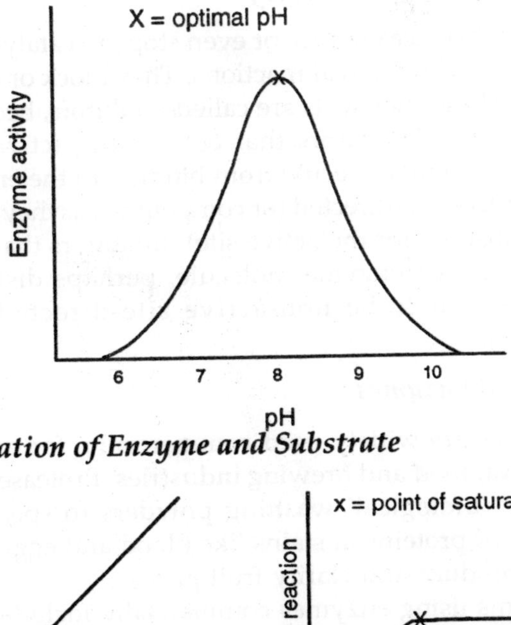

Concentration of Enzyme and Substrate

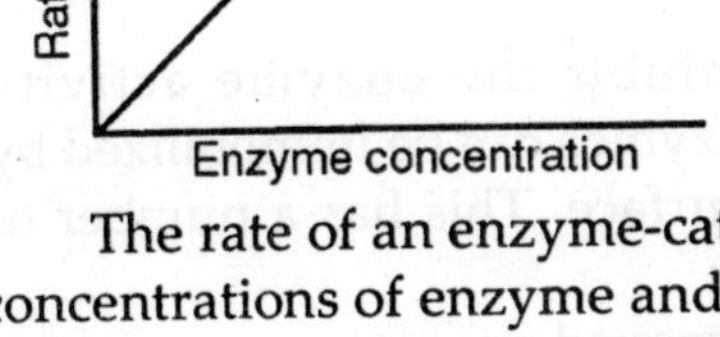

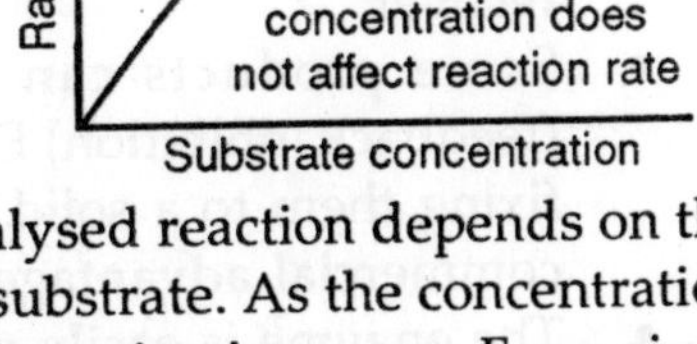

The rate of an enzyme-catalysed reaction depends on the concentrations of enzyme and substrate. As the concentration of either is increased the rate of reaction increases. For a given enzyme concentration, the rate of reaction increases with increasing substrate concentration up to a point, above which any further increase in substrate concentration produces no significant change in reaction rate.

This is because the active sites of the enzyme molecules at any given moment are virtually saturated with substrate. The enzyme/substrate complex has to dissociate before the active sites are free to accommodate more substrate Provided that the substrate concentration is high and that temperature and pH are kept constant, the rate of reaction is proportional to the enzyme concentration.

Inhibition of Enzyme activity

Some substances reduce or even stop the catalytic activity of enzymes in biochemical reactions. They block or distort the active site. These chemicals are called inhibitors, because they inhibit reaction. Inhibitors that occupy the active site and prevent a substrate molecule from binding to the enzyme are said to be active site-directed (or competitive, as they 'compete' with the substrate for the active site). Inhibitors that attach to other parts of the enzyme molecule, perhaps distorting its shape, are said to be non-active site-directed (or non competitive).

Immobilized Enzymes

Enzymes are widely used commercially, for example in the detergent, food and brewing industries. Protease enzymes are used in 'biological' washing powders to speed up the breakdown of proteins in stains like blood and egg. Pectinase is used to produce and clarify fruit juices.

Problems using enzymes commercially include:

- They are water soluble which makes them hard to recover
- Some products can inhibit the enzyme activity (feedback inhibition) Enzymes can be immobilized by fixing them to a solid surface. This has a number of commercial advantages:
- The enzyme is easily removed
- The enzyme can be packed into columns and used over a long period

- Speedy separation of products reduces feedback inhibition
- Thermal stability is increased allowing higher temperatures to be used
- Higher operating temperatures increase rate of reaction

There are four principal methods of immobilization currently in use:

- Covalent bonding to a solid support
- Adsorption onto an insoluble substance
- Entrapment within a gel
- Encapsulation behind a selectively permeable membrane

WORKING PROCESS OF ENZYMES

ACTIVE SITES AND CATALYTIC MECHANISMS

Open any textbook of biochemistry and you will be presented with an overwhelming number of figures depicting the crystal structures of a multitude of enzymes. Of course, the three-dimensional structures of enzymes are crucial to their functions.

Many enzymes are just regular protein molecules, composed of nothing else than the 20 standard amino acids. If you mix of all these amino acids in free form at, say, 10 mM each, this mixture will not have any significant catalytic activity. It therefore is the precise arrangement of the amino acids in the enzyme molecule that brings about the function.

As the á-amino and á-carboxyl groups of the amino acids are hooked up to each other in the polypeptide chain, they usually do not directly contribute to the catalytic effect of the enzyme. Instead, it is the side chains that are directly engaged win the reaction.

A very good example of this is chymotrypsin. Chymotrypsin is one of the major proteases in the human digestive tract, where its job is to knock down large protein molecules into small peptides that are then further processed by peptidases.

How does chymotrypsin do that? The enzyme-catalyzed reaction is similar to alkaline hydrolysis. In alkaline hydrolysis, a hydroxide ion, which is a strong nucleophile, attacks the carbon in the peptide bond that carries a partial positive charge. In the enzyme-catalyzed reaction, a deprotonated serine residue of the enzyme plays a role similar to that of the hydroxide ion.

Now, you know that serine normally is a neutral amino acid – its OH-group does not spontaneously dissociate, no more than the-OH group of alcohol does. The question therefore is, how does the enzyme deprotonate its own serine side chain? This is brought about by placing the serine next to a histidine and an aspartate inside the active centre. Aspartate deprotonates histidine, which in turn deprotonates the serine residue.

This motive – asp, his, ser – is very widespread among proteases and esterases, so much so that it is commonly called 'the catalytic triad'. E.g., the protease trypsin and several lipases that occur in human metabolism have this motif and share the same mechanism of catalysis.

Structure and Mechanism of Chymotrypsin:

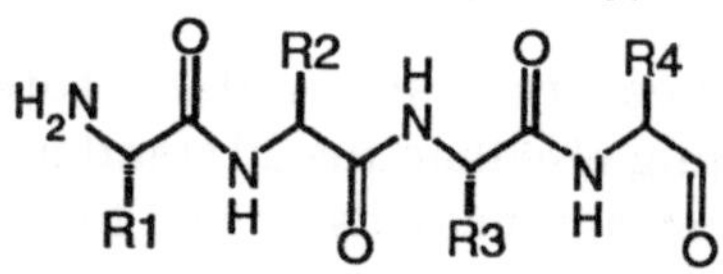

Fig. The Peptide Bond cleaved by chymotrypsin.

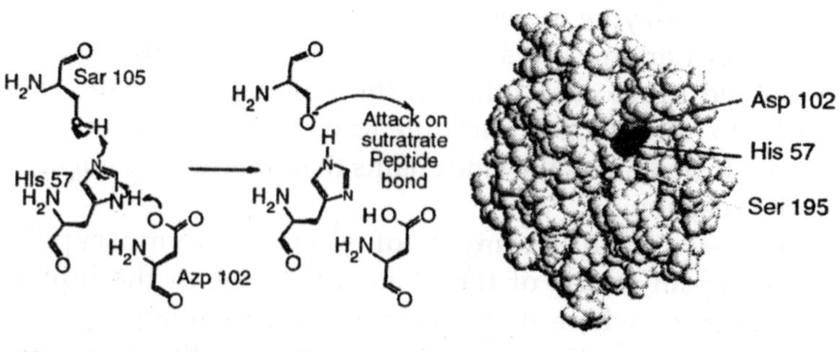

Fig. The Interplay of the Side Chains of Histidine 57, Aspartate 102, and Serine 195 that Results in the Deprotonation of Serine.

All three side chains are close to each other in the active site of the enzyme. While with many enzymes the protein molecule and its amino acid side chains are sufficient for catalysis, many others require co-enzymes for their catalytic activity. Very often, both active-site amino acid side chains and coenzymes are required.

For example, amino acid transaminases have a molecule of the coenzyme pyridoxalphosphate bound to the active site, which co-operates with an active site lysine in the enzyme's catalytic activity.

Most enzymes have just one active site, or if they are multimeric one active site per subunit. However, there are exceptions: Fatty acid synthase has as many as eight different active sites on each subunit. Multi-enzyme complexes such as pyruvate dehydrogenase have one active site per subunit but combine different types of subunits and enzyme activities in one functional assembly.

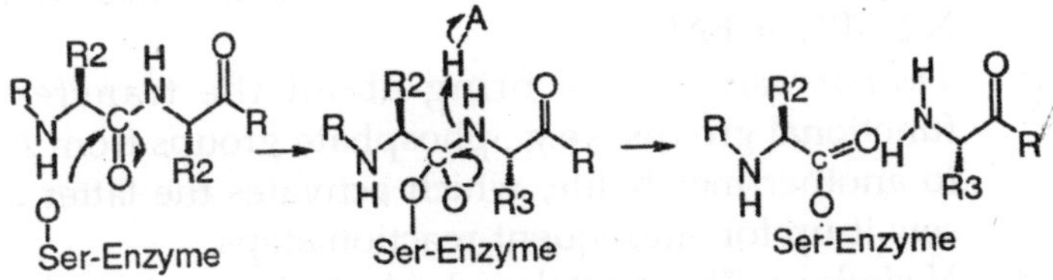

Fig. The Enzyme-Catalized reaction.

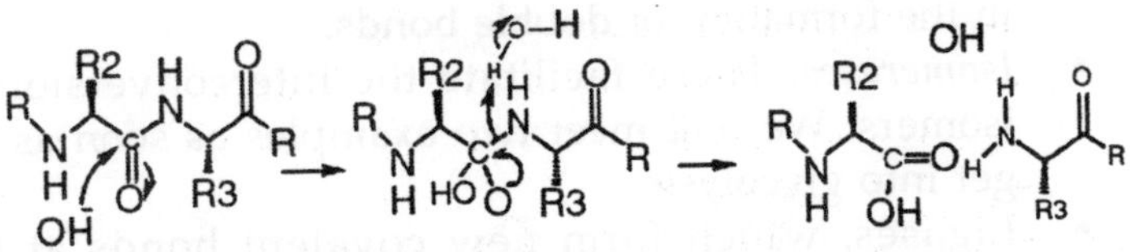

Fig. The Base-Catalyzed Reaction for Comparison.

TYPES OF ENZYMATIC REACTIONS

When looking at enzyme names such as 'transketolase' or 'phosphorylase', you will note that these names don't tell you exactly what reactions the enzymes may catalyze. A complete description should mention the coenzymes required, the substrates and the particular bonds in the substrates that are being severed or created.

A nomenclature that meets these criteria has been developed by the Enzyme Commission of the IUBMB (International Union of Biochemistry and Molecular Biology). In the IUBMB nomenclature, the enzyme transketolase bears the formidable name:

Sedoheptulose-7-phosphate: D-glyceraldehyde-3-phosphate glycolaldehydetransferase. Such names, of course, are rather lengthy, and their use is not very widespread. To make the tasks of tracking and bookkeeping more manageable, these names are supplemented with numeric codes. In the IUBMB scheme, enzymes are put into one out of six classes according to the reactions they catalyze.

These classes are:

- *Oxidoreductases*. These catalyze redox reactions, frequently involving one of the coenzymes NAD^+, $NADP^+$, or FAD.
- *Transferases*. These bring about the transfer of functional groups – e.g., phosphate groups from ATP to another metabolite, which activates the latter and sets it up for subsequent reaction steps.
- *Hydrolases*. These catalyze hydrolysis reactions – e.g., such as those involved in digestion of foodstuffs.
- *Lyases* – these effect elimination reactions that result in the formation of double bonds.
- *Isomerases*. These facilitate the interconversion of isomers. We will meet two examples as soon as we get into glycolysis.
- Ligases, which form new covalent bonds at the expense of ATP hydrolysis.

Of course, within each of these main classes, there are subclasses and sub-sub classes that correspond to details of substrates and mechanisms of the enzyme reactions.

Each individual enzyme activity is assigned an individual number within a sub-sub class, so that we wind up with a four-figure designation, which is preceded by the letters 'EC'

(Enzyme Commission). One good thing about this classification is that it is rarely used – the 'recommended names', which most of the time happen to be the traditional ones, are used instead. The other good thing is that it has a sound appreciation of priorities. The single most important enzyme in student lifestyle – namely, alcohol dehydrogenase (or, as IUBMB puts it, alcohol:NAD oxidoreductase). This beneficial enzyme, residing in the liver, degrades ethanol, and without it you would be drunk all the time!

Enzyme Kinetics: Basic Enzyme Reactions

Enzymes are catalysts and increase the speed of a chemical reaction without themselves undergoing any permanent chemical change. They are neither used up in the reaction nor do they appear as reaction products. The basic enzymatic reaction can be represented as follows where E represents the enzyme catalyzing the reaction, S the substrate, the substance being changed, and P the product of the reaction.

Enzyme Kinetics: Energy Levels

Chemists have known for almost a century that for most chemical reactions to proceed, some form of energy is needed. They have termed this quantity of energy, "the energy of activation." It is the magnitude of the activation energy which determines just how fast the reaction will proceed. It is believed that enzymes lower the activation energy for the reaction they are catalyzing. Figure illustrates this concept.

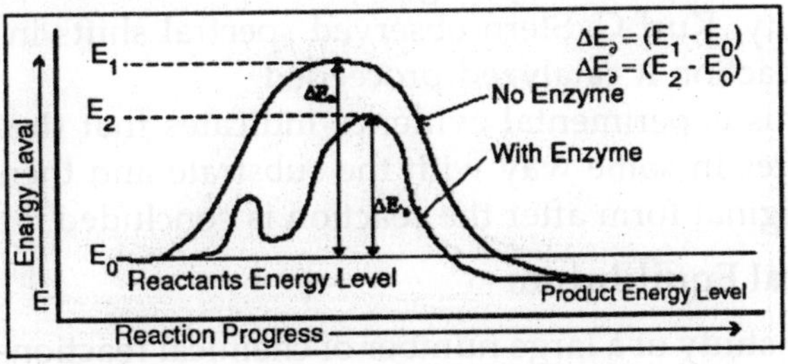

Fig. Energy of Activation Δ_∂ E is less than Δ_∂ E due to the Effect of the Enzyme on the Substrate.

The enzyme is thought to reduce the "path" of the reaction. This shortened path would require less energy for each molecule of substrate converted to product.

ZGiven a total amount of available energy, more molecules of substrate would be converted when the enzyme is present (the shortened "path") than when it is absent. Hence, the reaction is said to go faster in a given period of time.

Enzyme Kinetics: The Enzyme Substrate Complex

A theory to explain the catalytic action of enzymes was proposed by the Swedish chemist Savante Arrhenius in 1888. He proposed that the substrate and enzyme formed some intermediate substance which is known as the enzyme substrate complex.

The reaction can be represented as:

$$\underset{\text{substrate}}{S} + \underset{\text{enzyme}}{E} \longrightarrow \underset{\text{enzyme substrate complex}}{ES}$$

If this reaction is combined with the original reaction equation, the following results:

$$\underset{\text{substrate}}{S} + \underset{\text{enzyme}}{E} \longrightarrow \underset{\text{enzyme substrate complex}}{ES} \longrightarrow \underset{\text{Product}}{P} + \underset{\text{enzyme}}{E}$$

The existence of an intermediate enzyme-substrate complex has been demonstrated in the laboratory, for example, using catalase and a hydrogen peroxide derivative. At Yale University, Kurt G. Stern observed spectral shifts in catalase as the reaction it catalyzed proceeded.

ZThis experimental evidence indicates that the enzyme first unites in some way with the substrate and then returns to its original form after the reaction is concluded.

Chemical Equilibrium

The study of a large number of chemical reactions reveals that most do not go to true completion. This is likewise true of enzymatically-catalyzed reactions.

This is due to the reversibility of most reactions.

In general:

$$A + B \xrightarrow{K+1} C + D \quad \text{forward reaction}$$

$$C + D \xrightarrow{K-1} A + B \quad \text{reverse reaction}$$

where K^{+1} is the forward reaction rate constant and K^{-1} is the rate constant for the reverse reaction.

Combining the two reactions gives:

$$A + B \underset{K-1}{\overset{K+1}{\rightleftarrows}} C + D$$

Applying this general relationship to enzymatic reactions allows the equation:

$$E + S \underset{K-1}{\overset{K+1}{\rightleftarrows}} ES \underset{K-2}{\overset{K+2}{\rightleftarrows}} p + E$$

Equilbrium, a steady state condition, is reached when the forward reaction rates equal the backward rates. This is the basic equation upon which most enzyme activity studies are based.

Factors Affecting Enzyme Activity

Knowledge of basic enzyme kinetic theory is important in enzyme analysis in order both to understand the basic enzymatic mechanism and to select a method for enzyme analysis.

The conditions selected to measure the activity of an enzyme would not be the same as those selected to measure the concentration of its substrate.

Several factors affect the rate at which enzymatic reactions proceed-temperature, pH, enzyme concentration, substrate concentration, and the presence of any inhibitors or activators.

Enzyme Concentration

In order to study the effect of increasing the enzyme concentration upon the reaction rate,the substrate must be present in an excess amount; i.e., the reaction must be independent of the substrate concentration. Any change in the amount of product formed over a specified period of time will be dependent upon the level of enzyme present. These reactions are said to be "zero order" because the rates are independent of substrate concentration, and are equal to some constant k. The formation of product proceeds at a rate which is linear with time. The addition of more substrate does,

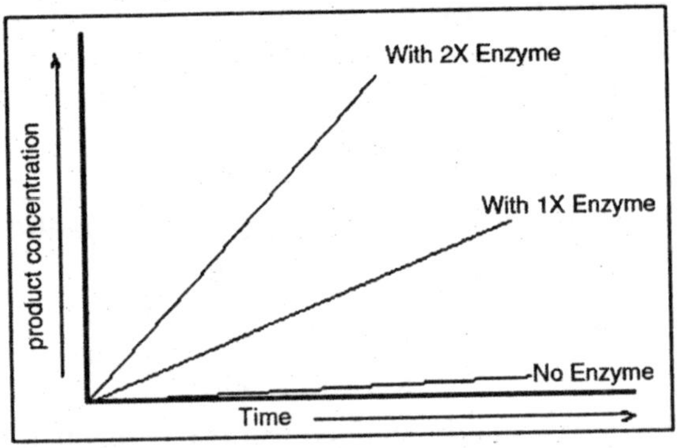

Fig. "Zer order" Reaction Rate is Independent of Substrate Concentration.

not serve to increase the rate. In zero order kinetics, allowing the assay to run for double time results in double the amount of product.

Table. Reaction Orders with Respect to Substrate Concentration

Order	Rate Equation	Comments
Zero	rate = k	rate is independent of substrate concentration
First	rate = k[S]	rate is proportional to the first power of substrate concentration
Second	rate = $k[S][S]=k[S]^2$	rate is proportional to the square of the substrate concentration
Second	rate = $k[S_1][S_2]$	rate is proportional to the first power of each of two reactants

The amount of enzyme present in a reaction is measured by the activity it catalyzes. The relationship between activity and concentration is affected by many factors such as temperature, pH, etc.

An enzyme assay must be designed so that the observed activity is proportional to the amount of enzyme present in order that the enzyme concentration is the only limiting factor.

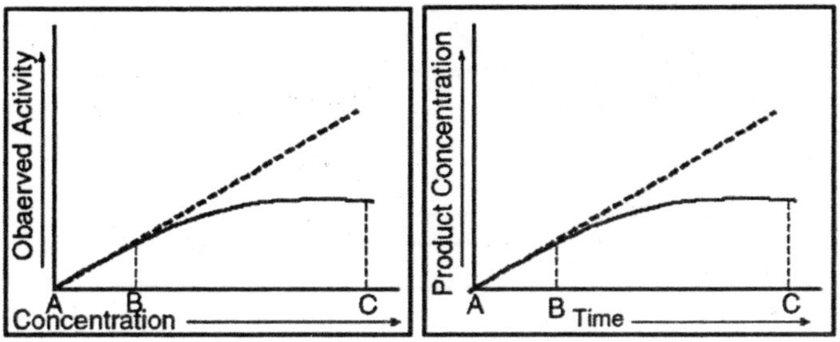

Fig. (a) Activity Concentration. (b) Reaction rate limited by substrate concentration

It is satisfied only when the reaction is zero order. In Figure, activity is directly proportional to concentration in the area AB, but not in BC.

Enzyme activity is generally greatest when substrate concentration is unlimiting. When the concentration of the product of an enzymatic Reaction is plotted against time, a similar curve results.

Between A and B, the curve represents a zero order reaction; that is, one in which the rate is constant with time. As substrate is used up, the enzyme's active sites are no longer saturated, substrate concentration becomes rate limiting, and the reaction becomes first order between B and C. To measure enzyme activity ideally, the measurements must be made in that portion of the curve where the reaction is zero order.

A reaction is most likely to be zero order initially since substrate concentration is then highest.

To be certain that a reaction is zero order, multiple measurements of product (or substrate) concentration must be made. Three types of reactions which might be encountered in enzyme assays and shows the problems which might be enountered if only single measurements are made.

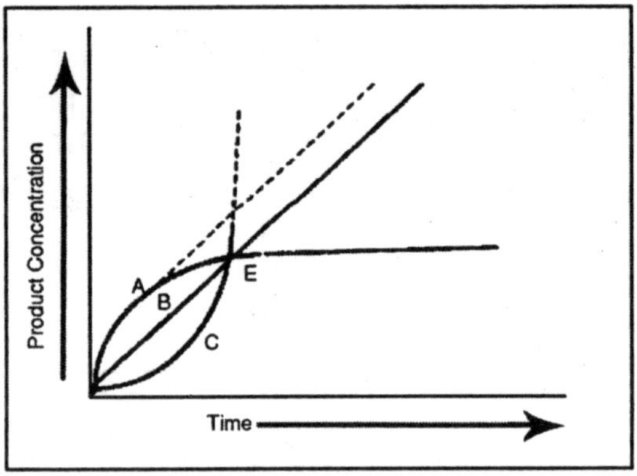

Fig. Leading, Lagging and Liner Reactions

B is a straight line representing a zero order reaction which permits accurate determination of enzyme activity for part or all of the reaction time.

This reaction is zero order initially and then slows, presumably due to substrate exhaustion or product inhibition. This type of reaction is sometimes referred to as a "leading" reaction.

True "potential" activity is represented by the dotted line. Curve C represents a reaction with an initial "lag" phase. Again the dotted line represents the potentially measurable activity.

Multiple determinations of product concentration enable each curve to be plotted and true activity determined. A single end point determination at E would lead to the false conclusion that all three samples had identical enzyme concentration.

Substrate Concentration

It has been shown experimentally that if the amount of the enzyme is kept constant and the substrate concentration is then gradually increased, the reaction velocity will increase until it reaches a maximum. After this point, increases in substrate concentration will not increase the velocity (Δ A/ Δ).

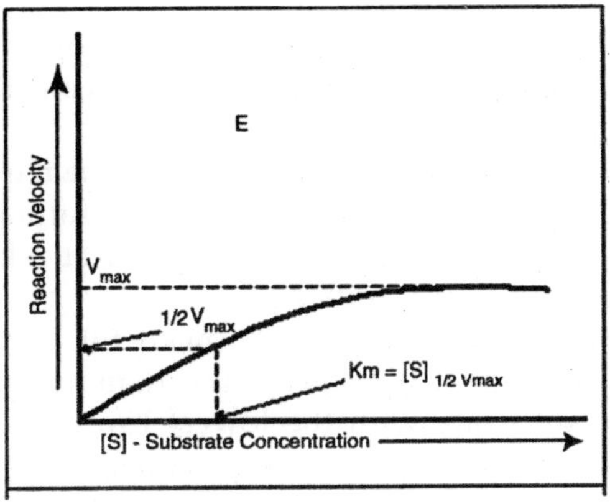

Fig. Effect of SConcentration

It is theorized that when this maximum velocity had been reached, all of the available enzyme has been converted to ES, the enzyme substrate complex. This point on the graph is designated V_{max}. Using this maximum velocity and equation, Michaelis developed a set of mathematical expressions to calculate enzyme activity in terms of reaction speed from measurable laboratory data.

$$E + S \underset{K-1}{\overset{K+1}{\rightleftharpoons}} ES \underset{K-2}{\overset{K+2}{\rightleftharpoons}} p + E$$

The Michaelis constant Km is defined as the substrate concentration at 1/2 the maximum velocity. Using this constant and the fact that Km can also be defined as:

$$K_m = K_{-1} + K_{2/}K_{+1}$$

K^{+1}, K^{-1} and K^{+2} being the rate constants from equation. Michaelis developed the following:

$$V_1 \frac{V_{mx}[S]}{K_m + [S]}$$

V_1 = The velocity at any time

[S] = The substrate concentration at this time

V_{mx} = The highest under this set of experimental condition (pH, temperature, etc.)

K_M = The Michaselis constant for the particular enzyme being investigated Michaelis constants have been determined for many of the commonly used enzymes. The size of Km tells us several things about a particular enzyme.

- A small Km indicates that the enzyme requires only a small amount of substrate to become saturated. Hence, the maximum velocity is reached at relatively low substrate concentrations.
- A large Km indicates the need for high substrate concentrations to achieve maximum reaction velocity.
- The substrate with the lowest Km upon which the enzyme acts as a catalyst is frequently assumed to be enzyme's natural substrate, though this is not true for all enzymes.

Effects of Inhibitors on Enzyme Activity

Enzyme inhibitors are substances which alter the catalytic action of the enzyme and consequently slow down, or in some cases, stop catalysis. There are three common types of enzyme inhibition-competitive, non-competitive and substrate inhibition. Most theories concerning inhibition mechanisms are based on the existence of the enzyme-substrate complex ES. As mentioned earlier, the existence of temporary ES structures has been verified in the laboratory. Competitive inhibition occurs when the substrate and a substance resembling the substrate are both added to the enzyme. A theory called the "lock-key theory" of enzyme catalysts can be used to explain why inhibition occurs.

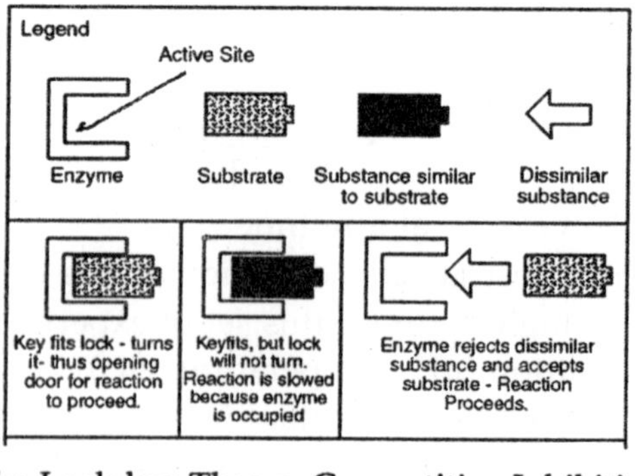

Fig. Lock-key Theory-Competitive Inhibition

The lock and key theory utilizes the concept of an "active site." The concept holds that one particular portion of the enzyme surface has a strong affinity for the substrate. The substrate is held in such a way that its conversion to the reaction products is more favorable.

If we consider the enzyme as the lock and the substrate the key-the key is inserted in the lock, is turned, and the door is opened and the reaction proceeds.

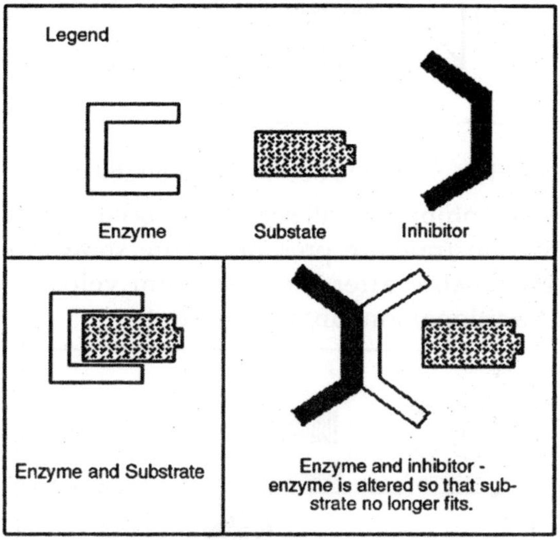

Fig. Non-Competitive Inhibition

However, when an inhibitor which resembles the substrate is present, it will compete with the substrate for the position in the enzyme lock.

When the inhibitor wins, it gains the lock position but is unable to open the lock. Hence, the observed reaction is slowed down because some of the available enzyme sites are occupied by the inhibitor.

If a dissimilar substance which does not fit the site is present, the enzyme rejects it, accepts the substrate, and the reaction proceeds normally.

Non-competitive inhibitors are considered to be substances which when added to the enzyme alter the enzyme in a way that it cannot accept the substrate.

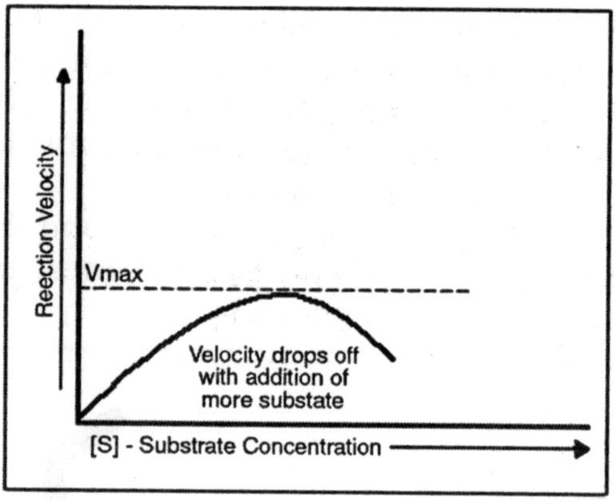

Fig. Substrate Become Rate-inhibiting

Substrate inhibition will sometimes occur when excessive amounts of substrate are present. Figure shows the reaction velocity decreasing after the maximum velocity has been reached. Additional amounts of substrate

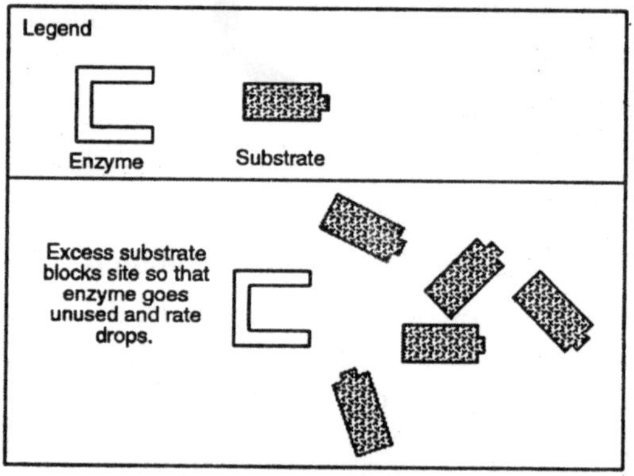

Fig. Substrate Inhibition

added to the reaction mixture after this point actually decrease the reaction rate. This is thought to be due to the fact that there are so many substrate molecules competing for the active sites on the enzyme surfaces that they block the sites and prevent any other substrate molecules from occupying them. This causes the reaction rate to drop since all of the enzyme present is not being used.

Temperature Effects

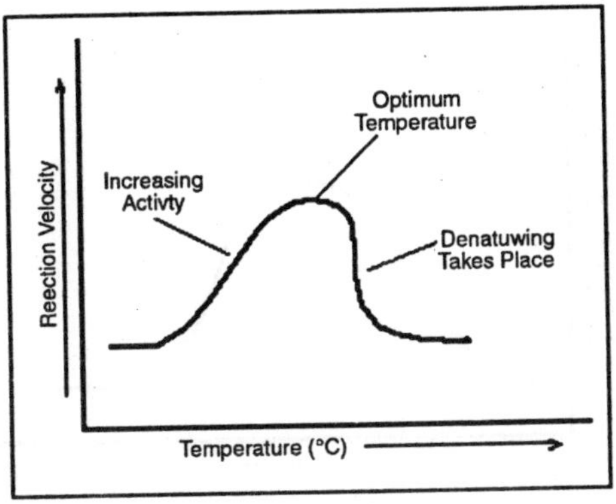

Fig. Effect of Temperature on Reaction Rate

Like most chemical reactions, the rate of an enzyme-catalyzed reaction increases as the temperature is raised. A ten degree Centigrade rise in temperature will increase the activity of most enzymes by 50 to 100%.

Variations in reaction temperature as small as 1 or 2 degrees may introduce changes of 10 to 20% in the results. In the case of enzymatic reactions, this is complicated by the fact that many enzymes are adversely affected by high temperatures.

The reaction rate increases with temperature to a maximum level, then abruptly declines with further increase of temperature.

Because most animal enzymes rapidly become denatured at temperatures above 40°C, most enzyme determinations are carried out somewhat below that temperature. Over a period of time, enzymes will be deactivated at even moderate temperatures. Storage of enzymes at 5°C or below is generally the most suitable. Some enzymes lose their activity when frozen.

Effects of pH

Enzymes are affected by changes in pH. The most favorable pH value-the point where the enzyme is most active-is known as the optimum pH.

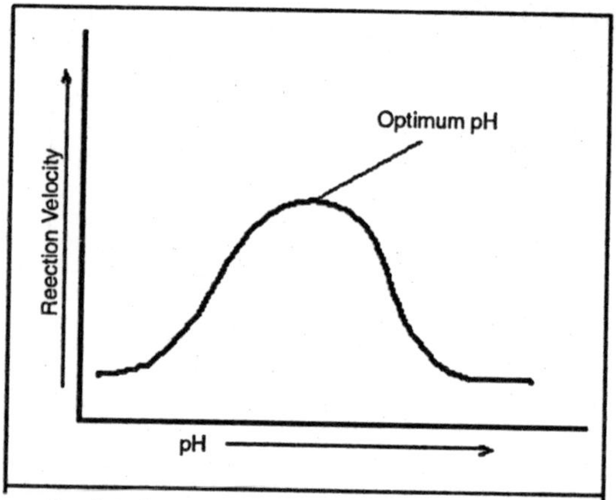

Extremely high or low pH values generally result in complete loss of activity for most enzymes. pH is also a factor in the stability of enzymes. As with activity, for each enzyme there is also a region of pH optimal stability.

The optimum pH value will vary greatly from one enzyme to another, table shows:

Table. pH for Optimum Activity.

Enzyme	pH Optimum
Lipase (pancreas)	8.0
Lipase (stomach)	4.0-5.0
Lipase (castor oil)	4.7
Pepsin	1.5-1.6
Trypsin	7.8-8.7
Urease	7.0
Invertase	4.5
Maltase	6.1-6.8
Amylase (pancreas)	6.7-7.0
Amylase (malt)	4.6-5.2
Catalase	7.0

In addition to temperature and pH there are other factors, such as ionic strength, which can affect the enzymatic reaction. Each of these physical and chemical parameters must be considered and optimized in order for an enzymatic reaction to be accurate and reproducible.

The term enzyme kinetics implies a study of the speed,

rate or velocity of an enzyme catalysed reaction, and of the various factors which may affect this.

At the heart of any study of enzyme kinetics is a knowledge of the way in which reaction velocity is altered by changes in the concentration of the enzyme's substrate and of the simple mathematics underlying this.

To ease ourselves gently into this we will assume that the enzyme that we are discussing has no special features, such as allosteric properties, and catalyses the conversion of just one substrate to one product.

This may seem unrelated to real enzymes, very few have just one substrate after all, but it will provide a basis which we can expand on later when we study more complex systems.

Enzyme kinetics is the study of the chemical reactions that are catalysed by enzymes, with a focus on their reaction rates. The study of an enzyme's kinetics reveals the catalytic mechanism of this enzyme, its role in metabolism, how its activity is controlled, and how a drug or a poison might inhibit the enzyme.

Enzymes are usually protein molecules that manipulate other molecules — the enzymes' substrates. These target molecules bind to an enzyme's active site and are transformed into products through a series of steps known as the enzymatic mechanism.

These mechanisms can be divided into single-substrate and multiple-substrate mechanisms. Kinetic studies on enzymes that only bind one substrate, such as triosephosphate isomerase, aim to measure the affinity with which the enzyme binds this substrate and the turnover rate.

When enzymes bind multiple substrates, such as dihydrofolate reductase (shown right), enzyme kinetics can also show the sequence in which these substrates bind and the sequence in which products are released.

An example of enzymes that bind a single substrate and release multiple products are proteases, which cleave one protein substrate into two polypeptide products.

Others join two substrates together, such as DNA polymerase linking a nucleotide to DNA. Although these mechanisms are often a complex series of steps, there is typically one rate-determining step that determines the overall kinetics.

This rate-determining step may be a chemical reaction or a conformational change of the enzyme or substrates, such as those involved in the release of product(s) from the enzyme.

Knowledge of the enzyme's structure is helpful in interpreting the kinetic data. For example, the structure can suggest how substrates and products bind during catalysis; what changes occur during the reaction; and even the role of particular amino acid residues in the mechanism.

Some enzymes change shape significantly during the mechanism; in such cases, it is helpful to determine the enzyme structure with and without bound substrate analogs that do not undergo the enzymatic reaction.

Not all biological catalysts are protein enzymes; RNA-based catalysts such as ribozymes and ribosomes are essential to many cellular functions, such as RNA splicing and translation.

The main difference between ribozymes and enzymes is that the RNA catalysts perform a more limited set of reactions, although their reaction mechanisms and kinetics can be analysed and classified by the same methods.

The reaction catalysed by an enzyme uses exactly the same reactants and produces exactly the same products as the uncatalysed reaction. Like other catalysts, enzymes do not alter the position of equilibrium between substrates and products.

However, unlike normal chemical reactions, enzymes are saturable. This means as more substrate is added, the reaction rate will increase, because more active sites become occupied. This can continue until all the enzyme becomes saturated with substrate and the rate reaches a maximum.

The two most important kinetic properties of an enzyme are how quickly the enzyme becomes saturated with a particular substrate, and the maximum rate it can achieve.

Knowing these properties suggests what an enzyme might do in the cell and can show how the enzyme will respond to changes in these conditions.

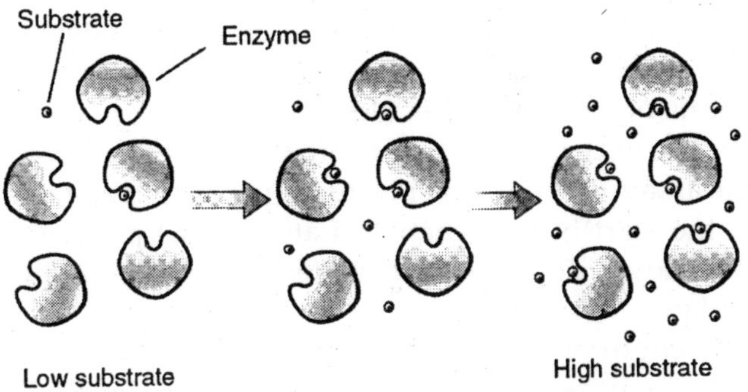

Fig. Reaction rates Increase as Substrate Concentration Increase, but Become Saturated at very High Concentrations of Substrate.

Enzyme assays are laboratory procedures that measure the rate of enzyme reactions. Because enzymes are not consumed by the reactions they catalyse, enzyme assays usually follow changes in the concentration of either substrates or products to measure the rate of reaction.

There are many methods of measurement. Spectrophotometric assays observe change in the absorbance of light between products and reactants; radiometric assays involve the incorporation or release of radioactivity to measure the amount of product made over time. Spectrophotometric assays are most convenient since they allow the rate of the reaction to be measured continuously.

Although radiometric assays require the removal and counting of samples (i. e., they are discontinuous assays) they are usually extremely sensitive and can measure very low levels of enzyme activity. An analogous approach is to use mass spectrometry to monitor the incorporation or release of stable isotopes as substrate is converted into product. The most sensitive enzyme assays use lasers focused through a microscope to observe changes in single enzyme molecules as they catalyse their reactions.

These measurements either use changes in the fluorescence of cofactors during an enzyme's reaction mechanism, or of fluorescent dyes added onto specific sites of the protein to report movements that occur during catalysis.

These studies are providing a new view of the kinetics and dynamics of single enzymes, as opposed to traditional enzyme kinetics, which observes the average behaviour of populations of millions of enzyme molecules.

On the left is shown a typical progress curve for an enzyme assay. The enzyme produces product at a linear initial rate at the start of the reaction. Later in this progress curve, the rate slows down as substrate is used up or products accumulate.

The length of the initial rate period depends on the assay conditions and can range from milliseconds to hours. Enzyme assays are usually set up to produce an initial rate lasting over a minute, to make measurements easier. However, equipment for rapidly mixing liquids allows fast kinetic measurements on initial rates of less than one second. These very rapid assays are essential for measuring pre-steady-state kinetics.

Most enzyme kinetics studies concentrate on this initial, linear part of enzyme reactions. However, it is also possible to measure the complete reaction curve and fit this data to a non-linear rate equation. This way of measuring enzyme reactions is called progress-curve analysis. This approach is useful as an alternative to rapid kinetics when the initial rate is too fast to measure accurately.

MICHAELIS-MENTEN EQUATION

Study of the impact made on the rate of an enzyme-catalysed reaction by changes in experimental conditions is known as enzyme kinetics. Knowledge of kinetics can be a very useful tool in understanding the mechanism by which an enzyme carries out its catalytic activity.

The effect of substrate concentration on the initial rate of an enzyme-catalysed reaction is a central concept in enzyme kinetics. When data are generated from experiments of this type and the results plotted as a graph of initial rate (v, y-axis) against substrate concentration ([S], x-axis), many enzymes exhibit a rectangular hyperbolic curve like the one shown in the diagram below.

Note: the use of square brackets, as for [S] above is shorthand notation for "concentration of S", a convention that will be used extensively in the derivation below.

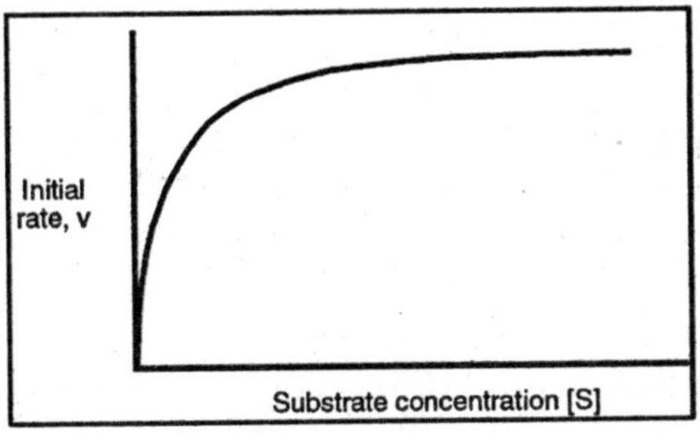

Observations of this type set Leonor Michaelis and Maud Menten thinking about the underlying reasons why a curve should follow this shape and led them to derive an algebraic equation that now bears their names. There are several modern ways to explain the way in which the Michaelis-Menten equation is derived, and one is spelt out below.

DERIVING THE MICHAELIS-MENTEN EQUATION

Start with the generalised scheme for enzyme-catalysed production of a product (P) from substrate (S). The enzyme (E) does not magically convert S into P, it must first come into physical contact with it, i. e. E binds S to form an enzyme-substrate complex (ES). Michaelis and Menten therefore set out the following scheme:

$$E + S \underset{k_1}{\overset{k_1}{\rightleftharpoons}} ES \xrightarrow{k_2} E + P$$

The terms k_1, k_{-1} and k_2 are rate constants for, respectively, the association of substrate and enzyme, the dissociation of unaltered substrate from the enzyme and the dissociation of product (= altered substrate) from the enzyme.

Note that there is the theoretical possibility of a reverse reaction, with ES complex forming from E and P, but this can be ignored because we are considering *initial* rates of reaction, i. e. when the enzyme is first provided with substrate, so there should not be any product available to combine with enzyme.

The overall rate of the reaction (v) is limited by the step ES to E + P, and this will depend on two factors-the rate of that step (i. e. k_2) and the concentration of enzyme that has substrate bound, i. e. [ES].

This can be written as:

$$v = k_2 [ES]$$

At this point it is important to draw your attention to two assumptions that are made in this scheme. The first is the availability of a vast excess of substrate, so that [S]>>[E]. Secondly, it is assumed that the system is in steady-state, i. e. that the ES complex is being formed and broken down at the same rate, so that overall [ES] is constant. The formation of ES will depend on the rate constant K_1 and the availability of enzyme and substrate, i. e. [E] and [S]. The breakdown of [ES] can occur in two ways, either the conversion of substrate to product or the non-reactive dissociation of substrate from the complex. In both instances the [ES] will be significant.

Thus, at steady state we can write:

$$K_1[E][S] = k_{-1}[ES]+K_2[ES]$$

The next couple of steps are rearrangements of this equation. First of all we can collect together the rate constants on the right-hand side because they are both multiplied by [ES], this gives us:

$$K_1[E][S] = k_{-1} + k_2[ES]$$

Then dividing both sides by (k_{-1} + k_2), this becomes:

$$\frac{k_1[E][S]}{(k_{-1}+k_2)} = [ES]$$

Notice that the three rate constants are now on the same side of the equation. As the name implies, these terms are constants, so we can actually combine them into one term. This new constant is termed the Michaelis constant and is written:

$$K_M.\frac{(k_{-1}+k_2)}{k_1} = k_M$$

Notice that the three rate constants in the definition of K_M are actually inverted (the other way up) compared with our previous equation. This is a 'trick' that makes for easier calculation at a later stage. Substituting this definition of K_M into our previous equation now gives us:

$$\frac{[E][S]}{K_M} = [ES]$$

The total amount of enzyme in the system must be the same throughout the experiment, but it can either be free (unbound) E or in complex with substrate, ES. If we term the total enzyme E_0, this relationship can be written out:

$$[E_0] = [E] + [ES]$$

This can be rearranged (by subtracting [ES] from each side) to give:

$$[E] = [E_0] - [ES]$$

So, the [E] free in solution is equal to the total amount of enzyme minus the amount that has substrate bound. Substituting this definition of [E] back into equation gives us:

$$\frac{([E_0] - [ES])[S]}{K_M} = [ES]$$

This can now be rearranged in several steps. First of all, open the bracket so that the terms $[E_0]$ and [ES] are separately multiplied by:

$$[S]\frac{[E_0][S] - [ES][S]}{K_M} = [ES]$$

Next, multiply each side by K_M, this gives us:

$$[E_0][S] - [ES][S] = k_M[ES]$$

Then collect the two [ES] terms together on the same side (you can either think of this as adding [ES][S] to both sides or as 'carry over and change the sign'-your preference will probably be an indication of how long ago you went to school).

This gives:

$$[E_0][S] = K_M[ES] + [ES][S]$$

Then because both terms on the right-hand side are multiplied by [ES] we can collect them together into a bracket:

$$[E_0][S] = (K_M + [S])[ES]$$

Dividing both sides by (K_M + [S]) now gives us:

$$\frac{[E_0][S]}{[S]+K_M}=[ES]$$

Substituting this left-hand side into Equation 1 in place of [ES] results in:

$$v=K_2\frac{[E_0][S]}{[S]+K_M}$$

The maximum rate, which we can call V_{max}, would be achieved when all of the enzyme molecules have substrate bound. Under conditions when [S] is much greater than [E], it is fair to assume that all E will be in the form ES. Therefore $[E_0]$ = [ES]. Thinking again about Equation, we could substitute the term V_{max} for v and $[E_0]$ for [ES].

This would give us:

$$V_{max} = k_2 [E_0]$$

Notice that $k_2[E_0]$ was present in our previous equation, so we can replace this with Vmax, giving a final equation:

$$v=\frac{v_{max}[S]}{k_M+[S]}$$

This final equation is actually called the Michaelis-Menten equation.

Perhaps this derivation still leaves you puzzled about the importance of the Michaelis-Menten equation. The significance becomes clearer when you consider the case when the rate of reaction (v) is exactly half of the maximal reaction rate (V_{max}). Under those circumstances, the Michaelis-Menten equation could be written:

$$\frac{v_{max}}{2}=\frac{v_{max}[S]}{k_M+[S]}$$

On dividing both sides by V_{max} this becomes:

$$\frac{1}{2}=\frac{[S]}{K_M+[S]}$$

Multiplying both sides by (KM + [S]) gives:

$$\frac{k_M+[S]}{2}=[S]$$

And then multiplying both sides by 2 further resolves the equation to:

$$k_M + [S] = 2[S]$$

2[S] on the right-hand side is the same as [S] + [S], so we can take away one [S] from each side. Thus when the rate of the reaction is half of the maximum rate:

$$k_M = [S]$$

The K_M of an enzyme is therefore the substrate concentration at which the reaction occurs at half of the maximum rate. If we now reconsider the graph that came at the start of this tutorial it could be written:

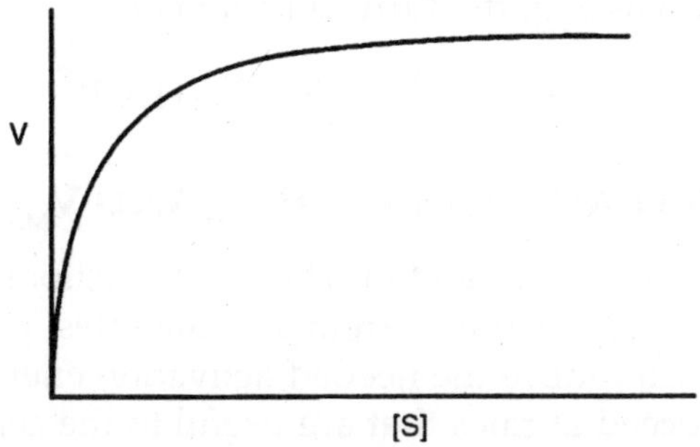

What does this all mean in physical terms? K_M is an indicator of the affinity that an enzyme has for a given substrate, and hence the stability of the enzyme-substrate complex.

Look at the shape of the graph. At low [S], it is the availability of substrate that is the limiting factor. Therefore as more substrate is added there is a rapid increase in the initial rate of the reaction-any substrate is rapidly mopped up and converted to product.

At the K_M, 50% of active sites have substrate bound. At higher [S] a point is reached (at least theoretically) where all of the enzyme has substrate bound and is working flat out. Adding more substrate will not increase the rate of the reaction, hence the levelling out observed in the graph. There are limitations in the quantitative (i. e. numerical) interpretation of this type of graph, known as a Michaelis plot.

The V_{max} is never really reached and therefore V_{max} and hence K_M values calculated from this graph are somewhat approximate.

A more accurate way to determine V_{max} and K_M (though still not perfect) is to convert the data into a linear Lineweaver-Burk plot.

"Experimental calculation of V_{max} and K_M "

If, incidentally, we consider that the rate constant k_2 for the conversion of ES to E + P in the initial scheme is the step determining the overall rate of production of P (as we have in deriving the Michaelis-Menten equation) then k_2 is actually the same term as k_{cat}, the turnover number.

$$E + S \underset{k_{-1}}{\overset{k_1}{\rightleftharpoons}} ES \xrightarrow{k_2} E + P$$

INTRACTEION WITH K_M AND V_{MAX}

Living systems depend on chemical reactions which, on their own, would occur at extremely slow rates. Enzymes are catalysts which reduce the needed activation energy so these reactions proceed at rates that are useful to the cell.

PRODUCT ACCUMULATION IS OFTEN LINEAR WITH TIME

In most cases, an enzyme converts one chemical (the *substrate*), into another (the *product*). A graph of product concentration vs. time follows three phases as shown in the following graph.

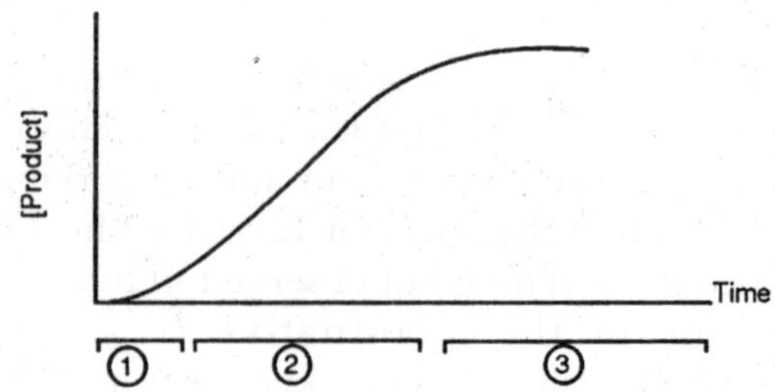

At very early time points, the rate of product accumulation increases over time. Special techniques are needed to study the early kinetics of enzyme action, since this transient phase usually lasts less than a second.

For an extended period of time, the product concentration increases linearly with time. At later times, the substrate is depleted, so the curve starts to level off. Eventually the concentration of product reaches a plateau and doesn't change with time.

It is difficult to fit a curve to a graph of product as a function of time, even if you use a simplified model that ignores the transient phase and assumes that the reaction is irreversible. The model simply cannot be solved to an equation that expresses product concentration as a function of time.

To fit these kind of data (called an *enzyme progress curve*) you need to use a programme that can fit data to a model defined by differential equations or by an implicit equation. Prism cannot do this.

Rather than fit the enzyme progress curve, most analyses of enzyme kinetics fit the initial velocity of the enzyme reaction as a function of substrate concentration.

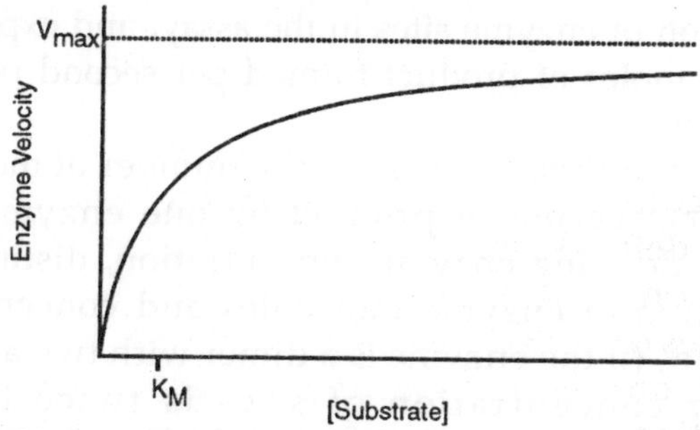

The velocity of the enzyme reaction is the slope of the linear phase, expressed as amount of product formed per time. If the initial transient phase is very short, you can simply measure product formed at a single time, and define the velocity to be the concentration divided by the time interval.

This chapter considers data collected only in the second phase. The terminology describing these phases can be confusing.

The second phase is often called the "initial rate", ignoring the short transient phase that precedes it. It is also called "steady state", because the concentration of enzyme-substrate complex doesn't change.

However, the concentration of product accumulates, so the system is not truly at steady state until, much later, the concentration of product truly doesn't change over time.

ENZYME VELOCITY AS A FUNCTION OF SUBSTRATE CONCENTRATION

If you measure enzyme velocity at many different concentrations of substrate, the graph generally looks like this: Enzyme velocity as a function of substrate concentration often follows the Michaelis-Menten equation:

$$\text{Velocity} = v = \frac{V_{max}[S]}{[S] + K_M}$$

Vmax is the limiting velocity as substrate concentrations get very large. Vmax (and V) are expressed in units of product formed per time. If you know the molar concentration of enzyme, you can divide the observed velocity by the concentration of enzyme sites in the assay, and express Vmax as units of moles of product formed per second per mole of enzyme sites.

This is the *turnover number,* the number of molecules of substrate converted to product by one enzyme site per second. In defining enzyme concentration, distinguish the concentration of enzyme molecules and concentration of enzyme sites (if the enzyme is a dimer with two active sites, the molar concentration of sites is twice the molar concentration of enzyme). KM is expressed in units of concentration, usually in Molar units. KM is the concentration of substrate that leads to half-maximal velocity. To prove this, set [S] equal to KM in the equation above. Cancel terms and you'll see that V=Vmax/2.

THE MEANING OF KM

To understand the meaning of Km, you need to have a

model of enzyme action. The simplest model is the classic model of Michaelis and Menten, which has proven useful with many kinds of enzymes.

$$E + S \underset{k-1}{\overset{k_1}{\rightleftharpoons}} ES \xrightarrow{K_2} E = P$$

The substrate (S) binds reversibly to the enzyme (E) in the first reaction. In most cases, you can't measure this step. What you measure is production of product (P), created by the second reaction.

From the model, we want to derive an equation that describes the rate of enzyme activity (amount of product formed per time interval) as a function of substrate concentration.

The rate of product formation equals the rate at which ES turns into E+P, which equals k2 times [ES]. This equation isn't helpful, because we don't know ES. We need to solve for ES in terms of the other quantities.

This calculation can be greatly simplified by making two reasonable assumptions. First, we assume that the concentration of ES is steady during the time intervals used for enzyme kinetic work.

That means that the rate of ES formation, equals the rate of ES dissociation (either back to E+S or forward to E+P). Second, we assume that the reverse reaction (formation of ES from E+P) is negligible, because we are working at early time points where the concentration of product is very low.

$$\text{Rate of ES formation} = \text{Rate of ES dissolution}$$

$$k_1.[S].[E_{free}] = k_{-1}[ES] + k_2.[ES]$$

We also know that the total concentration of enzyme, Etotal, equals ES plus E. So the equation can be rewritten.

Solving for ES:

$$[ES] = \frac{K_1.[E_{total}][S]}{k_1.[S] + k_2 + k_{-1}} = \frac{[E_{total}][S]}{[S] + \frac{k_2 + k_{-1}}{k_1}}$$

The velocity of the enzyme reaction therefore is:

$$\text{Velocity} = k_2[ES] = \frac{k_2[E_{total}][S]}{[S] + \frac{k_2 + k_{-1}}{k_1}}$$

Finally, define Vmax (the velocity at maximal concentrations of substrate) as k2 times Etotal, and KM, the Michaelis-Menten constant, as (k2+k-1)/k1.

Substituting:

$$\text{Velocity} = V = \frac{V_{max}[S]}{[S] + K_M}$$

Note that Km is not a binding constant that measures the strength of binding between the enzyme and substrate. Its value includes the affinity of substrate for enzyme, but also the rate at which the substrate bound to the enzyme is converted to product.

Only if k2 is much smaller than k-1 will KM equal a binding affinity. The Michaelis-Menten model is too simple for many purposes.

The Briggs-Haldane model has proven more useful:

$$E + S \rightleftarrows ES \rightleftarrows EP \longrightarrow E + P$$

Under the Briggs-Haldane model, the graph of enzyme velocity vs. substrate looks the same as under the Michaelis-Menten model, but KM is defined as a combination of all five of the rate constants in the model.

ASSUMPTIONS OF ENZYME KINETIC ANALYSES

Standard analyses of enzyme kinetics (the only kind discussed here) assume:

- The production of product is linear with time during the time interval used.
- The concentration of substrate vastly exceeds the concentration of enzyme. This means that the free concentration of substrate is very close to the concentration you added, and that substrate concentration is constant throughout the assay.
- A single enzyme forms the product.

- There is negligible spontaneous creation of product without enzyme
- No cooperativity. Binding of substrate to one enzyme binding site doesn't influence the affinity or activity of an adjacent site.
- Neither substrate nor product acts as an allosteric modulator to alter the enzyme velocity.

TO DETERMINE V_{MAX} AND K_M WITH PRISM

- Enter substrate concentrations into the X column and velocity into the Y column (entering replicates if you have them).
- Pick nonlinear regression from the list of curves and regressions.
- Choose more equations. Enter this equation as a new equation, or choose from the enzyme kinetics equation library.

$$Y = (V_{max} \times X)/(Km + X)$$

Variable	Comment
X	Substrate concentration. Usually expressed in µM or mM.
Y	Enzyme velocity in units of concentration of product per time. It is sometimes normalized to enzyme concentration, so concentration of product per time per concentration of enzyme.
Vmax	The maximum enzyme velocity. A reasonable rule for choosing an initial value might be that Vmax equals 1.0 times YMAX. Vmax is expressed in the same units as the Y values.
Km	The Michaelis-Menten constant. A reasonable rule for choosing an initial value might be 0.2*XMAX

COMPARISON OF ENZYME KINETICS WITH RADIOLIGAND BINDING

The Michaelis-Menten equation for enzyme activity has a form similar to the equation describing equilibrium binding.

$$\text{EnzymeVelocity} = V = \frac{V_{max}[S]}{[S]+K_M}$$

$$\text{SpecificBinding} = B = \frac{B_{max[L]}}{[L]+K_D}$$

Note these differences between binding experiments and enzyme kinetics.

- It usually takes many minutes or hours for a receptor incubation to equilibrate. It is common (and informative) to measure the kinetics prior to equilibrium. Enzyme assays reach steady state (defined as constant rate of product accumulation) typically in a few seconds. It is uncommon to measure the kinetics of the transient phase before that, although you can learn a lot by studying those transient kinetics.
- The equation used to analyse binding data is valid at equilibrium-when the rate of receptor-ligand complex formation equals the rate of dissociation. The equation used to analyse enzyme kinetic data is valid when the rate of product formation is constant, so product accumulates at a constant rate. But the overall system is not at equilibrium in enzyme reactions, as the concentration of product is continually increasing.
- KD is a dissociation constant that measures the strength of binding between receptor and ligand. KM is not a binding constant. Its value includes the affinity of substrate for enzyme, but also the kinetics by which the substrate bound to the enzyme is converted to product
- Bmax is measured as the number of binding sites normalized to amount of tissue, often fmol per milligram, or sites/cell. Vmax is measured as moles of product produced per minute.

DISPLAYING ENZYME KINETIC DATA ON A LINEWEAVER-BURK PLOT

The best way to analyse enzyme kinetic data is to fit the data directly to the Michaelis-Menten equation using nonlinear regression. Before nonlinear regression was available, investigators had to transform curved data into straight lines, so they could analyse with linear regression.

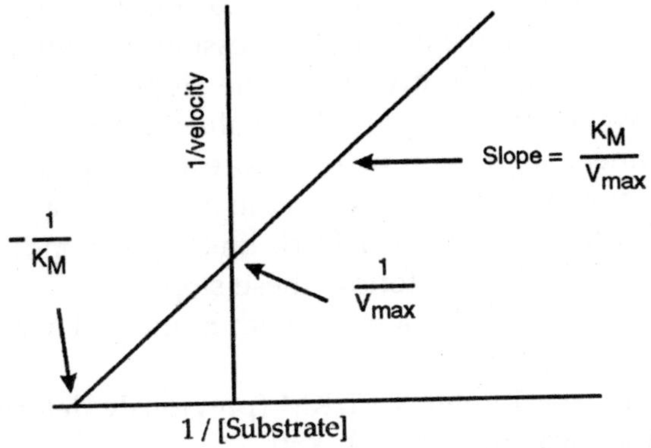

One way to do this is with a Lineweaver-Burk plot. Take the inverse of the Michaelis-Menten equation and simplify:

$$\frac{1}{v}=\frac{[S]+K_M}{V_{max}[S]}=\frac{[S]}{V_{max}[S]}+\frac{k_M}{V_{max}[S]}=\frac{1}{V_{max}}+\frac{1}{V_{max}}\cdot\frac{1}{[S]}$$

Ignoring experimental error, a plot of 1/V vs. 1/S will be linear, with a Y-intercept of $1/V_{max}$ and a slope equal to Km/V_{max}. The X-intercept equals $\simeq 1/Km$.

Use the Lineweaver-Burk plot only to display your data. Don't use the slope and intercept of a linear regression line to determine values for V_{max} and KM. If you do this, you won't get the most accurate values for V_{max} and KM. The problem is that the transformations (reciprocals) distort the experimental error, so the double-reciprocal plot does not obey the assumptions of linear regression. Use nonlinear regression to obtain the most accurate values of KM and V_{max}.

Tip. You should analyse enzyme kinetic data with nonlinear regression, not with Lineweaver-Burk plots. Use Lineweaver-Burk plots to display data, not to analyse data.

To create a Lineweaver-Burk plot with Prism, start from a table where X is substrate concentration and Y is velocity. Then choose Transformations from the list of data manipulations. Check the option boxes to transform both X to be 1/X, and Y to be 1/Y. Be sure to check the option to create a new graph of the results. From that graph, Analyse and choose linear regression to superimpose the regression line. This linear regression line should NOT be used to obtain values for V_{max} and Km. The X-intercept of the regression line will be near-1/KM, and the negative inverse of the slope will be near the V_{max}. However, the V_{max} and KM values determined directly with nonlinear regression will be more accurate. It is better to draw the line that corresponds to the nonlinear regression fit.

To create a Lineweaver-Burk line corresponding to the nonlinear regression fit, follow these steps:

- Create a new data table, with numerical X values and single Y values.
- Into row 1 enter X=-1/KM (previously determined by nonlinear regression), Y=0.
- Into row 2 enter X=1/S_{max} (S_{max} is the largest value of [substrate] you want to include on the graph) and Y=(1/V_{max})(1. 0 + KM/S_{max}).
- Note the name of this data table. Perhaps rename it to something appropriate.
- Go to the Lineweaver-Burk graph.
- Press Change, then Data on graph.
- Add the new data table to the graph.
- Press Change, then Symbols and lines.
- Drop down the list of data sets, and select the one you noted in Previous. Choose to plot no symbols, but to connect with a line.

ALLOSTERIC ENZYMES

One of the assumptions of Michaelis-Menten kinetics is that there is no cooperativity. If the enzyme is multimeric, then binding of a substrate to one binding site should have no effect on the activity of neighboring sites. This assumption is often not true. If binding of substrate to one binding site increases the activity of neighboring sites, the term *positive cooperativity* is used. Activity is related to substrate concentration by this equation:

$$\text{Velocity} = v = \frac{V_{max}[S]^h}{[S]^h + K_{0.5}^h}$$

When the variable h equals 1. 0, this equation is the same

as the Michaelis-Menten equation. With positive cooperativity, h will have a value greater than 1. 0.

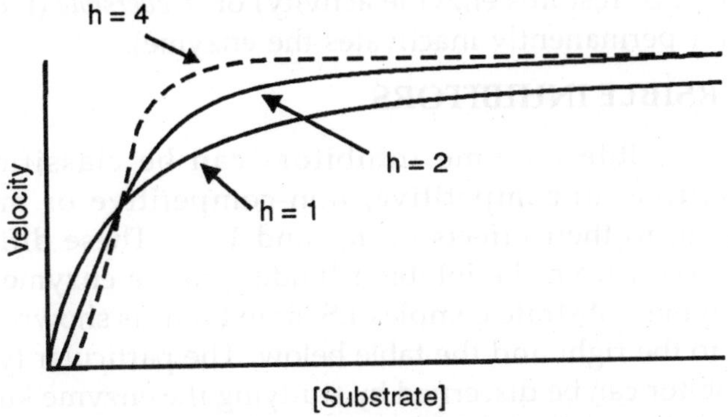

If there are two interacting binding sites, h will equal two (or less, depending on the strength of the cooperativity). If there are three interacting binding sites, h will equal 3 (or less). Note that the denominator has the new variable K0. 5 instead of KM. To fit data to the equation for enzyme activity with positive cooperativity, use the equation below. For initial values, try these rules:

V_{max}=Y_{max}, K=. 5 Xmid, and h=1. 0

Y=V_{max} Xh/(Kh + Xh)

The variable h does not always equal the number of interacting binding sites (although h can not exceed the number of interacting sites). Think of h as an empirical measure of the steepness of the curve and the presence of cooperativity.

ENZYME INHIBITION

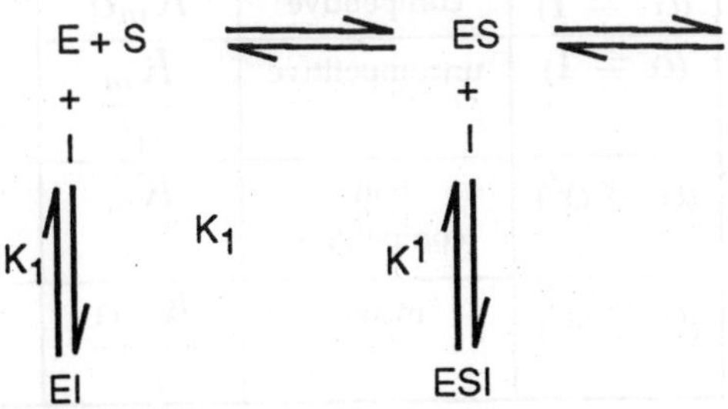

Fig. Kinetic Scheme for Reversible Enzyme Inhibitors.

Enzyme inhibitors are molecules that reduce or abolish enzyme activity. These are either *reversible* (i. e., removal of the inhibitor restores enzyme activity) or *irreversible* (i. e., the inhibitor permanently inactivates the enzyme).

REVERSIBLE INHIBITORS

Reversible enzyme inhibitors can be classified as competitive, uncompetitive, non-competitive or mixed, according to their effects on K_m and V_{max}. These different effects result from the inhibitor binding to the enzyme E, to the enzyme–substrate complex ES, or to both, as shown in the figure to the right and the table below. The particular type of an inhibitor can be discerned by studying the enzyme kinetics as a function of the inhibitor concentration.

The four types of inhibition produce Lineweaver–Burke and Eadie–Hofstee plots that vary in distinctive ways with inhibitor concentration.

For brevity, two symbols are used:

$$\alpha = 1 + \frac{[I]}{K_i} \quad \text{and} \quad \alpha' = 1 + \frac{[I]}{k_i'}$$

where K_i and K'_i are the dissociation constants for binding to the enzyme and to the enzyme–substrate complex, respectively. In the presence of the reversible inhibitor, the enzyme's apparent K_m and V_{max} become $(\alpha/\alpha')K_m$ and $(1/\alpha')V_{max}$, respectively, as shown below for common cases.

		Type of inhibition	K_m apparent	V_{max} apparent
K_i only	$(\alpha' = 1)$	competitive	$K_m\alpha$	V_{max}
K_i' only	$(\alpha = 1)$	uncompetitive	$\frac{K_m}{\alpha'}$	$\frac{V_{max}}{\alpha'}$
$K_i = K_i'$	$(\alpha = \alpha')$	non-competitive	K_m	$\frac{V_{max}}{\alpha'}$
$K_i \neq K_i'$	$(\alpha \neq \alpha')$	mixed	$\frac{K_m\alpha}{\alpha'}$	$\frac{V_{max}}{\alpha'}$

Non-linear regression fits of the enzyme kinetics data to

the rate equations above can yield accurate estimates of the dissociation constants K_i and K'_i.

IRREVERSIBLE INHIBITORS

Enzyme inhibitors can also irreversibly inactivate enzymes, usually by covalently modifying active site residues. These reactions, which may be called suicide substrates, follow exponential decay functions and are usually saturable. Below saturation, they follow first order kinetics with respect to inhibitor.

ENZYME KINETICS IN THE PRESENCE OF AN INHIBITOR

Competitive Inhibitors

If an inhibitor binds reversibly to the same site as the substrate, the inhibition will be competitive. Competitive inhibitors are common in nature. One way to measure the effect of an inhibitor is to measure enzyme velocity at a variety of substrate concentrations in the presence and absence of an inhibitor. As the graph below shows, the inhibitor substantially reduces enzyme velocity at low concentrations of substrate, but doesn't alter velocity very much at very high concentrations of substrate.

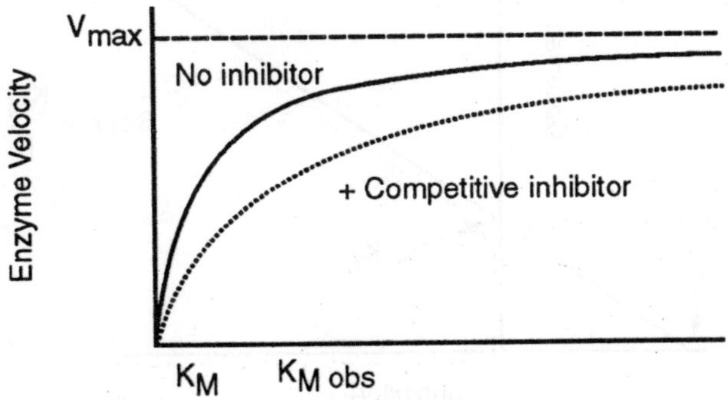

As the graph above shows, the inhibitor does not alter Vmax, but increases the observed KM (concentration of substrate that produces half-maximal velocity, in the presence of a competitive inhibitor). The observed KM is defined by the following equation, where Ki is the dissociation constant for inhibitor binding:

$$k_{M,obs} = k_M \left[1 + \frac{\text{Inhibitor}}{K_i}\right]$$

If you have determined the Km plus and minus a single concentration of inhibitor, you can rearrange that equation to determine the Ki.

$$k_i = \frac{[\text{Inhibitor}]}{\frac{K_{M,obs}}{K_M} - 1.0}$$

You'll get a more reliable determination of Ki if you determine the observed Km at a variety of concentrations of inhibitor. Fit each curve to determine the observed Km. Enter the results onto a new table, where X is the concentration of inhibitor, and Y is the observed Km.

If the inhibitor is competitive, the graph will be linear. Use linear regression to determine the X-and Y-intercepts. The Y-axis intercept equals the KM and the X-axis intercept equals the negative Ki.

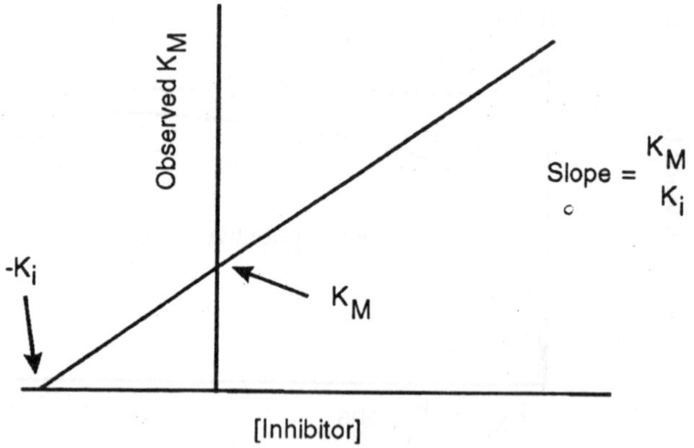

Another experimental design is to measure enzyme velocity at a single concentration of substrate with varying concentrations of a competitive inhibitor.

The results will look like this:

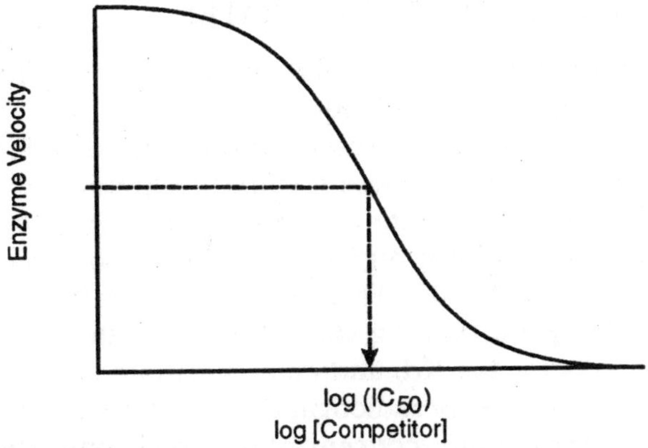

The concentration of competitor that reduces enzyme velocity by half is called the EC50 or IC50. Its value is determined by three factors:

- The dissociation constant for binding of inhibitor to enzyme, the Ki. If the Ki is low (the affinity is high), the EC50 will be low. The subscript i is used to indicate that the competitor inhibited enzyme activity. It is the concentration of the competitor that will bind to half the enzyme sites at equilibrium in the absence of substrate or other competitors.
- The concentration of the substrate. If you choose to use a higher concentration of substrate, it will take a larger concentration of inhibitor to compete for 50% of the activity.
- The KM. It takes more inhibitor to compete for a substrate with a low KM than for a substrate with a high KM.

Prism calculates the Ki, using the equation of Cheng and Prusoff:

$$k_i = \frac{IC_{50}}{1 + \frac{[\text{Substrate}]}{K_M}}$$

TO DETERMINE THE KI WITH PRISM (FROM DATA COLLECTED WITH A SINGLE CONCENTRATION OF SUBSTRATE)

- Enter data with X equal to logarithm of inhibitor concentration and Y equal to velocity.
- Press Analyse, and choose built-in analyses. From the curves section, choose nonlinear regression.
- Choose the one-site competitive binding equation.
- On the nonlinear regression dialog, choose the option "Ki from IC50". This option is usually used for radioligand binding studies, so enter the KM where it asks for the Kd, and enter substrate concentration where it asks for radioligand concentration. Enter both in mM (or any concentration units; only the ratio matters). Enter concentrations, not the logarithm of concentrations.

Inhibitors that are not Competitive

If an inhibitor binds to a site on the enzyme distinct from the site that binds substrate, the inhibition cannot be overcome by increasing the concentration of substrate. The inhibition is not competitive, and the inhibitor decreases the observed Vmax (and may also increase the observed KM). Competitive and non-competitive inhibitors bind reversibly. An inhibitor that binds covalently to irreversibly inactivate the enzyme is called an *irreversible inhibitor* or *inactivator*.

Chapter 9

Enzymes in Textiles

Biotechnology is the application of living organisms and their components to industrial products and processes. In 1981 thê European federation of biotechnology defined Biotechnology as. Integrated use of Biochemistry, Microbiology and Chemical Engineering in order to achieve the technological application of the capacities of microbes and cultured tissue cells.. The rapid developments in the field of genetic engineering have given a new impetus to biotechnology. This introduces the possibility of. tailoring. organisms in order to optimize the production of established or novel metabolites of commercial importance and of transferring genetic material (genes) from one organism to another.

Biotechnology also offers the potential for new industrial processes that require less energy and are based on renewable raw materials. It is important to note that biotechnology is not just concerned with biology, but it is a truly interdisciplinary subject involving the integration of natural and engineering sciences. Defining the scope of biotechnology is not easy because it overlaps with so many industries, such as, the chemical industry or food industry being the majors, but biotechnology has found many applications in textile industry also, especially in genetic engineering, textile processing and effluent management.

FARMING WITH BUGS

Simple cellular organisms, such as, yeast have been used for millennia, knowingly or otherwise, to make bread, beer and wine. Most industrial applications of biotechnology are still based upon fermentation processes using bacteria and enzymes to digest, transform and synthesize natural materials from one form into another. It is not surprising perhaps that biotechnology is often described as. farming with bugs.. For example, the active agent in many transformation processes is an enzyme rather than the cellular living organism itself. Enzymes are not alive themselves but are complex chemical catalysts, which can, in principle, be produced by a number of different methods, including non-biological synthetic routes.

THE SYSTEMATIC APPLICATION OF BIOLOGICAL SCIENCE

The contribution of science has been to understand to a much greater extent what exactly are the active components and mechanisms of the. Bugs. and their derivatives and therefore to begin to control, manipulate and reproduce their capabilities in a more systematic and intelligent fashion. Modern biotechnology has also brought forward a number of techniques, which do not rely on microbes and enzymes at all, but which directly modify the power of DNA molecule first and foremost among these is genetic engineering.

GENETIC ENGINEERING

With an improved understanding of how different genes are responsible for the various characteristics and properties of a living organism, techniques have been developed for isolating these active components (in particular, the DNA which carries the genetic code) and manipulating them outside of the cell. The next step has been to introduce fragments of DNA obtained from one organism into another, thereby transferring some of the properties and capabilities of the first to the second. For example, scientists working for the leading enzyme producer, Novo of Denmark, discovered that an enzyme produced in minute quantities by one particular fungus had very desirable properties for dissolving fats. The relevant genes where. spliced. into another micro-organism which was capable of producing the desired enzyme at much higher yield. Genetic engineering methods are being investigated for their potential to produce new kinds of textile fibres. The systems fall into two main groups. There are those systems that can produce monomeric protein molecules in solution from appropriately engineered genes and include expression in bacteria, cell cultures or in the milk of transgenic animals such as goats or sheep. The protein monomers are then isolated from the chosen system and spun and drawn into fibres.

The other approach is to modify keratin fibres such as wool by expressing other proteins in the internal components by trangenesis. Biopesticides based on a strain of soil bacteria known as Bt are already being used for control of caterpillar and beetle pests in a wide variety of fruits, vegetables and crops. More stable, longer lasting and more active Bts are now being developed for the suppression of loopers, bollworms and budworms in cotton.

The next stage will be to introduce greater insect and herbicide resistance by direct genetic engineering of the cotton plant itself. Practical results achieved so far include development of a cotton fibre with 50% greater strength than its. parent. Coloured cottons are also being developed, not only by conventional genetic selection but also by direct DNA engineering to produce, for example, deep blue cotton for denim production. The prospect is even being held out of encouraging natural polyesters such as polyhydroxy butyrate (PHB) to grow within the central hollow channel of the cotton fibre, thereby creating. natural. polyester-cotton.

MONOCLONAL ANTIBODIES

Monoclonal antibodies are protein molecules with an amazing ability to. recognize. specific substances, even at extremely low concentrations. They were first developed for use in medicine to detect and target cancer cells so called. magic bullet approach; they have also been used for pregnancy testing. Recently, Biocode Company has developed monoclonal antibodies as very sensitive marking tool for the prevention of counterfeiting. The markers themselves are cheap and safe substances, which can be applied to foodstuffs, drinks and textiles in concentrations of a few parts per million or less. The. codes. embodied in these markers are completely secure but can readily be detected by customs or trading standards inspectors using simple equipment in the field.

The technology has already been evaluated for the marking of branded denims. Methods have been perfected for use in nylon and acrylic resins and markers can also be incorporated into dyestuffs or applied to surfaces using ink jet printers.

DNA PROBES

DNA probes are another technology, which has grown out of genetic engineering. Short pieces of DNA can be designed to stick very specifically to other pieces of DNA and thereby, to help identify target species. The technique can be applied, for example, to distinguish Cashmere from Wool and other goat fibres. The initial impetus for application of DNA probes in the textile industry has come from importers and processors of specialty animal hairs who have seen a surge in trading and labelling fraud, especially in the wake of recent high fibre prices. Now, similar probes are being identified to distinguish between cotton, ramie, kapok, coir, flax, jute and hemp.

BIOSENSORS

Another way in which biological systems can be used as extremely sensitive analytical and control tools is biosensors. These employ some change produced by very small quantities of biologically active agents to measure and therefore, in principle, to control chemical and physical reactions. BTTG has been working on the use of certain fungi, which are capable of absorbing and concentrating heavy metal ions such as lead, copper and cadmium.

Resultant changes in the conductivity and dielectric properties of the fungi can be used to measure these species in a process or effluent stream relatively cheaply and easily. Application can be envisaged which incorporate biosensitive materials into textiles, for example, to produce. intelligent.

filter media or protective clothing which detects as well as protects against chemicals, gases and biological agents.

APPLICATIONS TO PROCESSING

The use of enzymes in textile processing and after care is already the best established example of the application of biotechnology to textiles and is likely to continue to provide some of the most immediate and possibly dramatic illustrations of its potential in the near-to medium-term future.

Fibre Preparation

Linen is a cellulosic fibre obtained from the flax plant. These fibres are formed in the cortex between the lignified core and the outer layers of the stem, they are separated from the stems by retting, in which matrix components, mainly pectin and lignin are removed and the fibres are separated. Recently, considerable efforts have been put to use enzymes in the retting process to control the process to produce linen fibres of consistent quality. Pre-treatment of the flax with sulphur dioxide gas brings about sufficient breakdown of the woody straw material to speed up enzyme retting whilst preventing excessive bacterial or fungal deterioration of the fibre. The carbonization process in which vegetable matter in wool is degraded by treatment with strong acid and then subjected to mechanical crushing can, in principle, be replaced by selective enzyme degradation of the impurities.

Fabric Preparation

Desizing using amylase enzymes has been well established for many years. However, there is still considerable scope for improving the speed, economics and consistency of the process, including the development of more temperature stable enzymes as well as a better understanding of how to characterize their activity and performance with respect to different fabrics, sizes, and processing conditions, eg, for pad batch as opposed to jigger desizing. The current application in the textile industry involves mainly hydrolases and now to some extent is Oxidoreductase. The Tables exemplify such textile applications. Another desirable development would be enzymes capable of destroying honeydew sugars, insect secretions that cause stickiness and severe processing problems for cotton spinners. An already established application is the use of catalase enzymes to breakdown residual hydrogen peroxide after, for example, pre-bleach of cotton that is to be dyed a pale or medium shade. Reactive dyes are especially sensitive to peroxide and currently require extended rinsing and/or use of chemical scavengers. The enzyme catalase is added after oxidative bleaching and allowed to react for 15 minutes at 30° C-40° C. It degrades the residual peroxide in water and oxygen.

Table. Application of Hydrolase Enzyme in Fabric Preparation

SI No.	Enzyme Name	Substrate Attacked	TextileApplication
1	Amylase	Starch	Starch desizing
2	Cellulase	Cellulose	1. Stone wash-Bio-polishing 2. Bio finishing for handle modification
3	Pectinase	Pectin	3. Carbonization of wool
4	Pectinase	Pectin	Bio scour replacing caustic
5	Catalase	Peroxides	In situ peroxide decomposition without any rinse in bleach bath
6	Proteases	Protein	1. Degumming of silk
	molecules	or peptide	2. Bioantifelting of wool bonds

7	Lipase	Fats and oils	Improve hydrophilicity of PET in place of alkaline hydrolysis.

Table. Application of Oxidoreductase in Fabric Preparation.

SI No.	Enzyme Name	Substrate Attacked	Textile Application
1	Laccase of coloured	Colour	1. Discoloration chromophore effluent and pigments 2. Bio-bleaching of lignin containing fibres like kenaf and jute 3. Bio-bleaching or indigo of in denim for various effects
2	Glucose	Pigments oxidases	In situ generation of H_2O_2 and bio-bleaching of cotton
3	AZO	Colour	Discoloration of AZO reductase chromophore dyes effluent and pigment
4	Peroxidase	colour ostreatus	Discoloration of chromophore

The results obtained were compared with the conventional process and it was found that the outcome of the enzymatic process was excellent. The best suitable conditions are the temperature range of 20° C-60° C, pH 5-10 and the application time is 10 min to 15 min.

FINISHING

Biostoning and the closely related process of bio-polishing are perhaps attracting most current attention in the area of enzyme processing. They are also an excellent illustration of how different industry structural and market considerations can affect the uptake of enzyme technology. Conventional stone washing uses abrasive pumice stones in a tumbling machine to abrade and remove particles of indigo dyestuff from the surfaces of denim yarns and fabric.

Cellulase enzymes can also cut through cotton fibres and achieve much the same effect without the damaging abrasion of the stones on both garment and machine. Disadvantages can include degradation of the fabric and loss of strength as well as. back staining. A slight reddening of the original indigo shade can also occur.

Now processors are learning to play more sophisticated tunes such as achieving a peach skin finish by use of a combination of stones and natural cellulase. Bio-polishing employs basically the same cellulose action to remove fine surface fuzz and fibrils from cotton and viscose fabrics.

The polishing action thus achieved helps to eliminate pilling and provides better print definition, colour brightness, surface texture, drapeability, and softness without any loss of absorbency. Bio-polishing can be used to clean up the fabric surface after the primary fibrillation of a peach skin treatment and prior to a secondary fibrillation process which imparts interesting fabric aesthetics.

A weight loss in the base fabric of some 3%-5% is typical but reduction in fabric strength can be controlled to within 2%-7% by terminating the treatment after about 30 min-40 min using a high temperature or low pH. enzyme stop.. One area that still poses problems is that of tubular cotton finishing.

Here, the fibre residues tend to be trapped inside the fabric rather than washed away.

WOOL PROCESSING APPLICATIONS

The international wool secretariat (IWS) together with, Novo, been developing the use of protease enzymes for a range of wool finishing treatments aimed at increased comfort (reduced prickle, greater softness) as well as improved surface appearance and pilling performance. The basic mechanisms closely parallel those of bio-polishing. The improved enzyme treatments will allow more selective removal of parts of the wool cuticle, there by modifying the luster, handle and felting characteristics without degradation or weakening of the wool fibre as a whole and without the need for environmentally damaging pre-chlorination treatment.

OTHER PROTEASE APPLICATIONS

Protease enzymes similar to those being developed for wool processing are already being used for the degumming of silk and for producing sand washed effects on silk garments. Treatment of Silk-Cellulosic blend is claimed to produce some unique effects. Proteases are also being used to wash down printing screens after use in order to remove the proteinaceous gums, which are used for thickening of printing pastes.

TEXTILE AFTER-CARE

Enzymes have been widely used in domestic laundering detergents since the 1960s. Some of the major classes of enzymes and their effectiveness against common stains are summarized in Table. Early problems of allergic reactions to some of these enzymes have now largely been overcome by the use of advanced granulation technology. Modern enzyme systems have reduced the use of sodium perborate in detergents by 25% along with the release of harmful salts into the environment.

However, enzymes still have to make a corresponding impact upon the commercial laundering market. One of the problems here has been the level of investment in. continuous-batch. or tunnel washers. These typically afford a residence time of 6 min-12 min which is not long enough for present enzyme systems to perform adequately. More efficient methods of. enzyme kill. are also required because of the extent of water recycling in modern washers.

Table. Types of Enzymes and their Effectiveness Against Various Stains

Enzyme	Effective for
Proteases	Grass, Blood, Egg, Sweat stains
Lipases	Lipstick, Butter, Salad oil, Sauces
Amylases	Spaghetti, Custard, chocloate
Celluloses	Colour brighterning, Softening, Soil removal

Further developments in the field of textile after-care may include treatments to reverse wool shrinkage as well as alternatives to dry cleaning.

CARING FOR THE ENVIRONMENT

Natural and enhanced microbial process have been used to treat waste materials and effluent streams from the textile industry. Conventional activated sludge and other systems are generally well able to meet BOD and related discharge limits on most cases. The industry faces some specific problems like colour removal from dyestuff effluent and handling of toxic wastes including PCPs and heavy metals. The synthetic dyes are designed in such a way that they become resistant to microbial degradation under the aerobic conditions. Also, the water solubility and the high molecular weight inhibit the permeation through biological cell membranes.

Anaerobic processes convert the organic contaminants principally into methane and carbon dioxide, usually occupy less space, treat wastes containing up to 30 000 mg/l of COD, have lower running costs and produce less sludge. A novel approach to promoting aerobic degradation in contaminated lagoons and preventing the development of malodorous and unpleasant anaerobic processes. The development based on a 3-D. biomat. of knitted polyester monofilament is used as a support for the micro-organisms. The mat is stable and resistant to compression; its open supporting structure counteracts the build-up of anaerobic sludges on the bottom of the lagoon.

NEW FIBRE SOURCES

Several possibilities exist for producing entirely new fibre materials, so called biopolymers, using biotechnological process routes, Naturally occurring polyester, PHB is produced by bacterial fermentation of a sugar feed stock and commercially available as. Biopol.. The polymer is stable under normal conditions but biodegrades completely in any microbially active environment. Other biopolymers with textile potential include polylactates and polycaprolactones, which are investigated for medical applications. Bacterial Cellulose The speciality papers and nonwovens are produced based on bacterially grown cellulose fibres these are extremely fine and resilient and are used as specialized filters, odour absorbers and reinforcing blends with aramids.

Genetically Modified Micro-Organisms Attempts have been made to transfer certain advantageous textile properties into micro-organisms where they can be more readily reproduced by bulk fermentation processes. The spider DNA is transferred into bacteria with the air of manufacturing proteins with the strength and resilience of spider silk for use in bulletproof vests.

DYESTUFFS AND INTER MEDIATES

Attempts have been made to synthesize bacterial forms of indigo as well as fungal pigments for use in the textile industry. Certain micro fungi are capable of yielding up to 30% of their biomass as pigment. Potential non-textile applications include food industry colorants This note of caution needs to be echoed across the whole spectrum of biotechnology developments. Although biological systems after many attractive possibilities and new approaches to all sorts of problems and needs, considerable advances are still being made in. conventional. technologies, such as, catalysis, chemical synthesis and physical fibre modification which need to be kept in perspective. There is also still great concern in society about the unbridled advance of biotechnology, especially with regard to the modification of natural species with possible unknown long-term consequences.

NATURE'S ENZYMES AND BIOTECHNOLOGIC INDUSTRIAL ENZYMES

Campo Research-JTC Enzymes Laboratories of Japan produces the same important types of enzyme as those found in malt and apples, for instance, as well as a wide range of novel enzymes specialities from various array of temperature and tropical fruits, for use in the cosmetic industry.

Campo's enzymes are just as natural, but they come from a different source. Almost all of Campo's cosmetic industrial enzymes originate from microorganisms' genetic extrapolated into fruit cells. Microorganisms can be bacteria, funds or yeasts and different species thrive in different conditions. The search for a new enzyme can start by examining water and soil samples natural products samples and botanical samples from the far corners of the world.

Just one microorganism can contain over 1, 000 different enzymes. A long period of trial and error in the laboratory is needed to isolate the best microorganism for producing a particular type of enzyme. When the right microorganism has been found, the work is still not over. The microorganism's genes has to be modified and transcripted into the biotechnologic fruit cells, so that it (micro-organisms genetic transcription combined in the biotechnologic fruits' genes) is capable of producing the desired enzyme at high yields in the biotechnologic fruit cells.

Then the fruit cells is 'grown' in huge mass-tissue culture vats or tanks where it produces the desired enzyme. The technique of mass-tissue culture fruit cells has made it possible to produce enzymes economically and in virtually unlimited quantities. The end-product of mass-tissue-culture is primed matured fruit cells in a broth from which the enzymes are extracted, After this, the remaining media broth is centrifuged or filtered to remove all solid particles. The resulting biomass, or sludge in everyday language, contains the residues of fruit cells and raw materials.

For past 25 years, tobacco in Japan has been produced in this way and now, this technology is used in converting this techniques to produce cosmetic enzymes such as fruit enzymes with detergency properties, and skin-care enzymes, sunscreen and UV-filter enzymes. The by-products (sludge)of this technology is disposed in the best way to minimize impact on the environment. A recycling project was started as early as 1963, converting a large part of the waste sludge into pulverized anhydrous powder for agriculture industry to be used as fertilizer. This is an example of industry giving back to nature, rather than depleting natural resources.

Nature, it seems, has provided all the enzymes the world is likely to need; literally millions, each with a specific role in nature. By isolating them and producing them on a large scale, they can be used in the service of mankind for a vast range of applications. This can be illustrated briefly by going back to the apple and malt mentioned earlier above, as an comparison:

Campo produces a range of sunscreen/solar UV-filtering enzymes called UV-zymes and UV-extremo-zymes,; Pseudo-Phyto-p53 enhancer that is of the same type as that found in all the various plants/herbs to protect themselves from solar UV-radiation. By adding this enzyme, a sunscreen topical formulation the SPF and UV filtering and absorbing properties can dramatically increase and the less or none of any other chemical or organic sunscreen are needed. In this way, industrial enzymes can be used to give nature a helping hand.

Many of the enzymes contained in fruit can also be produced bio-technologically via biotechnological techniques.. In fact, in some cases the industrial enzymes are an improvement on the fermentation-obtained enzymes because they are better adapted to the conditions found in industrial applications.

An other example is fruit enzymes with detergency properties, these used to replace the harsh irritant potential surfactants, and consequently these fruit enzyme can be replaced as detergents and its content, can vary considerably 0. 1-50% in a concentrate.

These are a few of the possibilities opened up by biotechnologic industrial enzymes. There are many more and the main applications of enzymes in industry are described under Industrial Applications.

Chapter 10

Antibiotics

Antibiotics are chemical substances that can inhibit the growth of, and even destroy, harmful microorganisms. They are derived from special microorganisms or other living systems, and are produced on an industrial scale using a fermentation process. Although the principles of antibiotic action were not discovered until the twentieth century, the first known use of antibiotics was by the Chinese over 2,500 years ago. Today, over 10,000 antibiotic substances have been reported. Currently, antibiotics represent a multibillion dollar industry that continues to grow each year.

Antibiotics are used in many forms—each of which imposes somewhat different manufacturing requirements. For bacterial infections on the skin surface, eye, or ear, an antibiotic may be applied as an ointment or cream. If the infection is internal, the antibiotic can be swallowed or injected directly into the body. In these cases, the antibiotic is delivered throughout the body by absorption into the bloodstream.

Antibiotics differ chemically so it is under-standable that they also differ in the types of infections they cure and the ways in which they cure them. Certain antibiotics destroy bacteria by affecting the structure of their cells. This can occur in one of two ways. First, the antibiotic can weaken the cell walls of the infectious bacteria, which causes them to burst. Second, antibiotics can cause the contents of the bacterial cells to leak out by damaging the cell membranes. Another way in which antibiotics function is by interfering with the bacteria's metabolism. Some antibiotics such as tetracycline and erythromycin interfere with protein synthesis. Antibiotics like rifampin inhibit nucleic acid biosynthesis. Still other antibiotics, such as sulfonamide or trimethoprim have a general blocking effect on cell metabolism.

The commercial development of an antibiotic is a long and costly proposal. It begins with basic research designed to identify organisms, which produce antibiotic compounds. During this phase, thousands of species are screened for any sign of antibacterial action. When one is found, the species is tested against a variety of known infectious bacteria. If the results are promising, the organism is grown on a large scale so the compound responsible for the antibiotic effect can be isolated.

This is a complex procedure because thousands of antibiotic materials have already been discovered. Often, scientists find that their new antibiotics are not unique. If the material passes this phase, further testing can be done. This typically involves clinical testing to prove that the antibiotic works in animals and humans and is not harmful. If these tests are passed, the Food and Drug Administration (FDA) must then approve the antibiotic as a new drug. This whole process can take many years. The large-scale production of an antibiotic depends on a fermentation process. During fermentation, large amounts of the antibiotic-producing organism are grown. During fermentation, the organisms produce the antibiotic material, which can then be isolated for use as a drug. For a new antibiotic to be economically feasible, manufacturers must be able to get a high yield of drug from the fermentation process, and be able to easily isolate it. Extensive research is usually required before a new antibiotic can be commercially scaled up.

While our scientific knowledge of antibiotics has only recently been developed, the practical application of antibiotics has existed for centuries. The first known use was by the Chinese about 2,500 years ago. During this time, they discovered that applying the moldy curd of soybeans to infections had certain therapeutic benefits. It was so effective that it became a standard treatment. Evidence suggests that other cultures used antibiotic-type substances as therapeutic agents. The Sudanese-Nubian civilization used a type of tetracycline antibiotic as early as 350 A.D. In Europe during the Middle Ages, crude plant extracts and cheese curds were also used to fight infection.

Although these cultures used antibiotics, the general principles of antibiotic action were not understood until the twentieth century. The development of modern antibiotics depended on a few key individuals who demonstrated to the world that materials derived from microorganisms could be used to cure infectious diseases. One of the first pioneers in this field was Louis Pasteur. In 1877, he and an associate discovered that the growth of disease-causing anthrax bacteria could be inhibited by a saprophytic bacteria.

They showed that large amounts of anthrax bacilli could be given to animals with no adverse affects as long as the saprophytic bacilli were also given. Over the next few years, other observations supported the fact that some bacterially derived materials could prevent the growth of disease-causing bacteria.

In 1928, Alexander Fleming made one of the most important contributions to the field of antibiotics. In an experiment, he found that a strain of green *Penicillium* mold inhibited the growth of bacteria on an agar plate. This led to the development of the first modern era antibiotic, penicillin. A few years later in 1932, a paper was published which suggested a method for treating infected wounds using a penicillin preparation. Although these early samples of penicillin were functional, they were not reliable and further refinements were needed. These improvements came in the early 1940s when Howard Florey and associates discovered a new strain of *Penicillium,* which produced high yields of penicillin. This allowed large-scale production of penicillin, which helped launch the modern antibiotics industry.

After the discovery of penicillin, other antibiotics were sought. In 1939, work began on the isolation of potential antibiotic products from the soil bacteria streptomyces. It was around this time that the term antibiotic was introduced. Selman Waxman and associates discovered streptomycin in

1944. Subsequent studies resulted in the discovery of a host of new, different antibiotics including actinomycin, streptothricin, and neomycin all produced by *Streptomyces*. Other antibiotics that have been discovered since include bacitracin, polymyxin, viomycin, chloramphenicol and tetracyclines. Since the 1970s, most new antibiotics have been synthetic modifications of naturally occurring antibiotics.

RAW MATERIALS

The compounds that make the fermentation broth are the primary raw materials required for antibiotic production. This broth is an aqueous solution made up of all of the ingredients necessary for the proliferation of the microorganisms. Typically, it contains a carbon source like molasses, or soy meal, both of which are made up of lactose and glucose sugars. These materials are needed as a food source for the organisms. Nitrogen is another necessary compound in the metabolic cycles of the organisms.

For this reason, an ammonia salt is typically used. Additionally, trace elements needed for the proper growth of the antibiotic-producing organisms are included. These are components such as phosphorus, sulfur, magnesium, zinc, iron, and copper introduced through water soluble salts. To prevent foaming during fermentation, anti-foaming agents such as lard oil, octadecanol, and silicones are used.

THE MANUFACTURING PROCESS

Although most antibiotics occur in nature, they are not normally available in the quantities necessary for large-scale production. For this reason, a fermentation process was developed. It involves isolating a desired microorganism, fueling growth of the culture and refining and isolating the final antibiotic product. It is important that sterile conditions be maintained throughout the manufacturing process, because contamination by foreign microbes will ruin the fermentation.

Starting the culture

- Before fermentation can begin, the desired antibiotic-producing organism must be isolated and its numbers must be increased by many times. To do this, a starter culture from a sample of previously isolated, cold-stored organisms is created in the lab. In order to grow the initial culture, a sample of the organism is transferred to an agar-containing plate. The initial culture is then put into shake flasks along with food and other nutrients necessary for growth. This creates a suspension, which can be transferred to seed tanks for further growth.

- The seed tanks are steel tanks designed to provide an ideal environment for growing microorganisms. They are filled with the all the things the specific microorganism would need to survive and thrive, including warm water and carbohydrate foods like lactose or glucose sugars. Additionally, they contain other necessary carbon sources, such as acetic acid, alcohols, or hydrocarbons, and nitrogen sources like ammonia salts. Growth factors like vitamins, amino acids, and minor nutrients round out the composition of the seed tank contents. The seed tanks are equipped with mixers, which keep the growth medium moving, and a pump to deliver sterilized, filtered air. After about 24-28 hours, the material in the seed tanks is transferred to the primary fermentation tanks.

Fermentation

- The fermentation tank is essentially a larger version of the steel, seed tank, which is able to hold about 30,000 gallons. It is filled with the same growth media found in the seed tank and also provides an environment inducive to growth. Here the microorganisms are allowed to grow and multiply. During this process, they excrete large quantities of the desired antibiotic. The tanks are cooled to keep the temperature between 73-81° F (23-27.2 ° C). It is constantly agitated, and a continuous stream of sterilized air is pumped into it. For this reason, anti-foaming agents are periodically added. Since pH control is vital for optimal growth, acids or bases are added to the tank as necessary.

Isolation and Purification

- After three to five days, the maximum amount of antibiotic will have been produced and the isolation process can begin. Depending on the specific antibiotic produced, the fermentation broth is processed by various purification methods. For example, for antibiotic compounds that are water soluble, an ion-exchange method may be used for purification. In this method, the compound is first separated from the waste organic materials in the broth and then sent through equipment, which separates the other water-soluble compounds from the desired one. To isolate an oil-soluble antibiotic such as penicillin, a solvent extraction method is used. In this method, the broth is treated with organic solvents such as butyl acetate or methyl isobutyl ketone, which can specifically dissolve the antibiotic. The dissolved antibiotic is then recovered using various organic chemical means. At the end of this step, the manufacturer is typically left with a purified powdered form of the antibiotic, which can be further refined into different product types.

Refining

- Antibiotic products can take on many different forms. They can be sold in solutions for intravenous bags or syringes, in pill or gel capsule form, or they may be sold as powders, which are incorporated into topical ointments. Depending on the final form of the antibiotic, various refining steps may be taken after the initial isolation. For intravenous bags, the crystalline antibiotic can be dissolved in a solution, put in the bag, which is then hermetically sealed. For gel capsules, the powdered antibiotic is physically filled into the bottom half of a capsule then the top half is mechanically put in place. When used in topical ointments, the antibiotic is mixed into the ointment.

- From this point, the antibiotic product is transported to the final packaging stations. Here, the products are stacked and put in boxes. They are loaded up on trucks and transported to various distributors, hospitals, and pharmacies. The entire process of fermentation, recovery, and processing can take anywhere from five to eight days.

QUALITY CONTROL

Quality control is of utmost importance in the production of antibiotics. Since it involves a fermentation process, steps must be taken to ensure that absolutely no contamination is introduced at any point during production. To this end, the medium and all of the processing equipment are thoroughly steam sterilized. During manufacturing, the quality of all the compounds is checked on a regular basis. Of particular importance are frequent checks of the condition of the microorganism culture during fermentation. These are accomplished using various chromatography techniques. Also, various physical and chemical properties of the finished product are checked such as pH, melting point, and moisture content.

In the United States, antibiotic production is highly regulated by the Food and Drug Administration (FDA). Depending on the application and type of antibiotic, more or less testing must be completed. For example, the FDA requires that for certain antibiotics each batch must be checked by them for effectiveness and purity. Only after they have certified the batch can it be sold for general consumption.

Since the development of a new drug is a costly proposition, pharmaceutical companies have done very little research in the last decade. However, an alarming development has spurred a revived interest in the development of new antibiotics. It turns out that some of the disease-causing bacteria have mutated and developed a resistance to many of the standard antibiotics. This could have grave consequences on the world's public health unless new antibiotics are discovered or improvements are made on the ones that are available. This challenging problem will be the focus of research for many years to come.

PRODUCTION OF ANTIBIOTICS

The mass production of antibiotics began during World War II with streptomycin and penicillin. Now most antibiotics are produced by staged fermentations in which strains of microorganisms producing high yields are grown under optimum conditions in nutrient media in fermentation tanks holding several thousand gallons. The mold is strained out of the fermentation broth, and then the antibiotic is removed from the broth by filtration, precipitation, and other separation methods. In some cases new antibiotics are laboratory synthesized, while many antibiotics are produced by chemically modifying natural substances; many such derivatives are more effective than the natural substances against infecting organisms or are better absorbed by the body, e.g., some semisynthetic penicillins are effective against bacteria resistant to the parent substance.

Antibiotic, any of a variety of substances, usually obtained from microorganisms, that inhibit the growth of or destroy certain other microorganisms.

TYPES OF ANTIBIOTICS

The great number of diverse antibiotics currently available can be classified in different ways, e.g., by their chemical structure, their microbial origin, or their mode of action. They are also frequently designated by their effective range. Tetracyclines, the most widely used broad-spectrum antibiotics, are effective against both Gram-positive and Gram-negative bacteria, as well as against rickettsias and psittacosis-causing organisms. Ciprofloxacin (Cipro) is another broad-spectrum antibiotic, effective in the treatment of mild infections of the urinary tract and sinuses.

The medium-spectrum antibiotics bacitracin, the erythromycins, penicillin, and the cephalosporins are effective primarily against Gram-positive bacteria, although the streptomycin group is effective against some Gram-negative and Gram-positive bacteria. Polymixins are narrow-spectrum antibiotics effective against only a few species of bacteria.

Antibiotics are either injected, given orally, or applied to the skin in ointment form. Many, while potent anti-infective agents, also cause toxic side effects. Some, like penicillin, are highly allergenic and can cause skin rashes, shock, and other manifestations of allergic sensitivity.

Others, such as the tetracyclines, cause major changes in the intestinal bacterial population and can result in superinfection by fungi and other microorganisms. Chloramphenicol, which is now restricted in use, produces severe blood diseases, and use of streptomycin can result in ear and kidney damage. Many antibiotics are less effective than formerly because antibiotic-resistant strains of microorganisms have emerged.

Antibiotics have found wide nonmedical use. Some are used in animal husbandry, along with vitamin B_{12}, to enhance the weight gain of livestock. However, some authorities believe the addition of antibiotics to animal feeds is dangerous because continuous low exposure to the antibiotic can sensitize humans to the drug and make them unable to take the substance later for the treatment of infection. In addition low levels of antibiotics in animal feed encourage the emergence of antibiotic-resistant strains of microorganisms.

Drug resistance has been shown to be carried by a genetic particle transmissible from one strain of microorganism to another, and the presence of low levels of antibiotics can actually cause an increase in the number of such particles in the bacterial population and increase the probability that such particles will be transferred to pathogenic, or disease-causing, strains. Antibiotics have also been used to treat plant diseases such as bacteria-caused infections in tomatoes, potatoes, and fruit trees. The substances are also used in experimental research.

Although for centuries preparations derived from living matter were applied to wounds to destroy infection, the fact that a microorganism is capable of destroying one of another species was not established until the latter half of the 19th cent. when Pasteur noted the antagonistic effect of other bacteria on the anthrax organism and pointed out that this action might be put to therapeutic use.

Meanwhile the German chemist Paul Ehrlich developed the idea of selective toxicity: that certain chemicals that would be toxic to some organisms, e.g., infectious bacteria, would be harmless to other organisms, e.g., humans.

In 1928, Sir Alexander Fleming, a Scottish biologist, observed that *Penicillium notatum,* a common mold, had destroyed staphylococcus bacteria in culture, and in 1939 the American microbiologist René Dubos demonstrated that a soil bacterium was capable of decomposing the starchlike capsule of the pneumococcus bacterium, without which the pneumococcus is harmless and does not cause pneumonia.

Dubos then found in the soil a microbe, *Bacillus brevis,* from which he obtained a product, tyrothricin, that was highly toxic to a wide range of bacteria.

Tyrothricin, a mixture of the two peptides gramicidin and tyrocidine, was also found to be toxic to red blood and reproductive cells in humans but could be used to good effect when applied as an ointment on body surfaces. Penicillin was finally isolated in 1939, and in 1944 Selman Waksman and Albert Schatz, American microbiologists, isolated streptomycin and a number of other antibiotics from *Streptomyces griseus.*

ETHANOL

Ethanol fermentation is the biological process by which sugars such as glucose, fructose, and sucrose are converted into cellular energy and thereby producing ethanol and carbon dioxide as metabolic waste products.

Yeasts carry out ethanol fermentation on sugars in the absence of oxygen. Because the process does not require oxygen, ethanol fermentation is classified as anaerobic. Ethanol fermentation is responsible for the rising of bread dough, the production of ethanol in alcoholic beverages, and for much of the production of ethanol for use as fuel.

ETHANOL

Ethanol (ethyl alcohol, grain alcohol) is a clear, colorless liquid with a characteristic, agreeable odor. In dilute aqueous solution, it has a somewhat sweet flavour, but in more concentrated solutions it has a burning taste. Ethanol, CH_3CH_2OH, is an alcohol, a group of chemical compounds whose molecules contain a hydroxyl group, –OH, bonded to a carbon atom. The word *alcohol* derives from Arabic *al-kuhul,* which denotes a fine powder of antimony used as an eye makeup. *Alcohol* originally referred to any fine powder, but medieval alchemists later applied the term to the refined products of distillation, and this led to the current usage.

Ethanol melts at –114.1°C, boils at 78.5°C, and has a density of 0.789 g/mL at 20°C. Its low freezing point has made it useful as the fluid in thermometers for temperatures below –40°C, the freezing point of mercury, and for other low-temperature purposes, such as for antifreeze in automobile radiators. Ethanol has been made since ancient times by the fermentation of sugars. All beverage ethanol and more than half of industrial ethanol is still made by this process. Simple sugars are the raw material. Zymase, an enzyme from yeast, changes the simple sugars into ethanol and carbon dioxide.

The fermentation reaction, represented by the simple equation:

$$C_6H_{12}O_6 \rightarrow 2CH_3CH_2OH + 2CO_2$$

Is actually very complex, and impure cultures of yeast produce varying amounts of other substances, including glycerine and various organic acids. In the production of beverages, such as whiskey and brandy, the impurities supply the flavour. Starches from potatoes, corn, wheat, and other plants can also be used in the production of ethanol by fermentation. However, the starches must first be broken down into simple sugars. An enzyme released by germinating barley, diastase, converts starches into sugars. Thus, the germination of barley, called malting, is the first step in brewing beer from starchy plants, such as corn and wheat. The ethanol produced by fermentation ranges in concentration from a few per cent up to about 14 per cent.

Above about 14 per cent, ethanol destroys the zymase enzyme and fermentation stops. Ethanol is normally concentrated by distillation of aqueous solutions, but the composition of the vapour from aqueous ethanol is 96 per cent ethanol and 4 per cent water. Therefore, pure ethanol cannot be obtained by distillation. Commercial ethanol contains 95 per cent by volume of ethanol and 5 per cent of water. Dehydrating agents can be used to remove the remaining water and produce absolute ethanol.

Much ethanol not intended for drinking is now made synthetically, either from acetaldehyde made from acetylene, or from ethylene made from petroleum. Ethanol can be oxidized to form first acetaldehyde and then acetic acid. It can be dehydrated to form ether. Butadiene, used in making synthetic rubber, may be made from ethanol, as can chloroform and many other organic chemicals. Ethanol is used as an automotive fuel by itself and can be mixed with gasoline to form gasohol.

Ethanol is miscible (mixable) in all proportions with water and with most organic solvents. It is useful as a solvent for many substances and in making perfumes, paints, lacquer, and explosives. Alcoholic solutions of nonvolatile substances are called tinctures; if the solute is volatile, the solution is called a spirit. Most industrial ethanol is denatured to prevent its use as a beverage. Denatured ethanol contains small amounts, 1 or 2 per cent each, of several different unpleasant or poisonous substances. The removal of all these substances would involve a series of treatments more expensive than the federal excise tax on alcoholic beverages (currently about $20 per gallon).

These denaturants render ethanol unfit for some industrial uses. In such industries undenatured ethanol is used under close federal supervision. When an alcoholic beverage is swallowed, it passes through the stomach into the small intestine where the ethanol is rapidly absorbed and distributed throughout the body. The ethanol enters body tissues in proportion to their water content. Therefore, more ethanol is found in the blood and the brain than in muscle or fat tissue. The ethanol is greatly diluted by body fluids. For example, a

1-ounce shot of 100-proof whiskey, which contains 0.5 fluid ounces of ethanol (about 15 mL), is diluted 5000-fold in a 150-pound human, producing a 0.02% blood alcohol concentration.

The chemical process of fermentation: The chemical equation below summarizes ethanol fermentation, in which one hexose molecule is converted into two ethanol molecules and two carbon dioxide molecules:

$$C_6H_{12}O_6 \rightarrow 2\ C_2H_5OH + 2\ CO_2$$

The process begins with a molecule of glucose being broken down by the process of glycolysis into pyruvate:

$$C_6H_{12}O_6 \rightarrow 2\ CH_3COCOO^- + 2H^+$$

This reaction is accompanied by the reduction of two molecules of NAD^+ to NADH and a net of two ADP molecules converted to two ATP plus the two water molecules.

Pyruvate is then converted to acetaldehyde and carbon dioxide. The acetaldehyde is subsequently reduced to ethanol by the NADH from the previous glycolysis, which is returned to NAD^+:

$$CH_3COCOO^- + H^+ \rightarrow CH_3CHO + CO_2$$
$$CH_3CHO + NADH \rightarrow C_2H_5OH + NAD^+$$

Yeast will perform the above two reactions only if oxygen is excluded from the environment. Otherwise yeast will oxidize pyruvate completely to carbon dioxide and water.

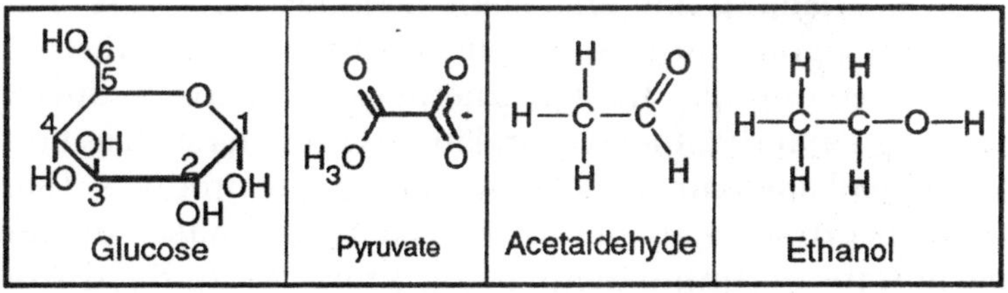

USES

Ethanol fermentation is responsible for the rising of bread dough. Yeast organisms consume sugars in the dough and produce ethanol and carbon dioxide as waste products. The carbon dioxide forms bubbles in the dough, expanding it into something of a foam. Nearly all the ethanol evaporates from the dough when the bread is baked.

The production of all alcoholic beverages employs ethanol fermentation by yeast. Wines and brandies are produced by fermentation of the natural sugars present in fruits, especially grapes. Beers, ales, and whiskeys employ fermentation of grain starches that have been converted to sugar by the application of the enzyme, amylase, which is present in grain kernels that have been germinated.

Amylase-treated grain or amylase-treated potatoes are fermented for the production of vodka. Fermentation of cane sugar is the first step in producing rum. In all cases, the fermentation must take place in a vessel that is arranged to allow carbon dioxide to escape, but that prevents outside air from coming in, as exposure to oxygen would prevent the formation of ethanol. Similar yeast fermentation of various carbohydrate products is used to produce much of the ethanol used for fuel.

SINGLE CELL PROTEIN

Single-cell proteins (SCP) refer to the dried cells of microorganisms. SCP's are used as protein sources in human foods or animal feeds. Many raw materials have been considered as carbon and energy sources for SCP production. In many cases, these raw materials have been hydrolyzed by physical, chemical and enzymatic methods before use. Ram horns are significant waste products of the meat industry in Turkey. For example, the slaughterhouse in Erzurum directly discharges about 25 tons a year. Fibrous proteins such as horn, feather, nail and hair are also abundant waste products. These waste products can be converted to biomass, protein concentrate or amino acids using proteases derived from certain microorganisms.

Horns consist of a-keratin, which has a very high content of cysteine and they also contain most of the common amino acids. Horns have the components of bone and blood tissues and are rich in some growth factors required by microorganisms. The main aim of this study was to investigate the suitability of horn hydrolysate as a substrate in SCP production.

SCP is the name given to a variety of microbial products, that are produced by fermentation. When properly produced, this materials make satisfactory proteinaceous ingredients for animal feed or human food. The production of protein from hydrocarbon wastes of the petroleum industry is the most recent microbiological industry.

7Yeast, fungi, bacteria, and algae are grown on hydrocarbon wastes, and cells are harvested as sources of protein. It has been calculated that 100 lbs of yeast will produce 250 tons of proteins in 24 hours, whereas a 1000 lbs steer will synthesize only 1 lb of protein 24 hours and this after consuming 12 to 20 lbs of plant proteins. Similar, algae grown in ponds can produce 20 tons (dry weight) of protein, per acre, per year. This yield is 10 to 15 times higher than soybean and 25 to 50 times higher than corn. There are both advantages and disadvantages in using microorganisms for animal or human consumption. Bacteria are usually high in protein (50 to 80 per cent) and have a rapid growth rate.

The principal disadvantages are as follows:

- Bacterial cells have small size and low density, which makes harvesting from the fermented medium difficult and costly.
- Bacterial cells have high nucleic acid content relative to yeast and fungi This can be detrimental to human beings, tending to increase the uric acid level in blood. This may cause uric acid poising or gout. To decrease the nucleic acid level additional processing step has to be introduced, and this increases the cost.
- The general public thinking is that all bacteria are harmful and produce disease. An extensive education programme is1required to remove this misconception and to make the public accept bacterial protein.

Yeasts have as advantages their larger size (easier to harvest), lower nucleic add content, high lysine content and ability to grow at acid pH. However the most important advantage is familiarity and acceptability because of the long history of its use in traditional fermentations. Disadvantages include lower growth rates, lower protein content (45 to 65 per cent), and lower methionine content than in bacteria. Filamentous fungi have advantages.in ease of harvesting, but have their limitations in lower growth rates, lower protein content, and acceptability. Algae have disadvantages of having cellulosic cell walls which are not digested by human beings. Secondly, they also concentrate heavy metals. Single cell protein basically comprises proteins, fats carbohydrates, ash ingredients, water, and other elements such as phosphorus and Potassium. The composition depends upon the organism and the substrate which it grows. some typical compositions which are compared with soymeal and fish meal. If SCP is to be used successfully, there are five main criteria to be satisfied;

- The SCP must be safe to eat.
- The nutritional value dependent on the amino acid composition must be high.
- It must be acceptable to the general public.
- It must have the functionality, i.e. characteristics, which are found in common staple foods.
- The economic viability of the SCP process is extremely complex and is yet to be demonstrated

BIOMASS AND SINGLE CELL PROTEIN

Microbial cells are produced for two main applications:

- As a source of protein for animal or human food,
- For use as a commercial inoculum in food fermentations and for agriculture and waste treatment.

As a commodity, SCP must be competitive with commercial animal and plant proteins, in terms of proce and nutritional value and must conform with human and animal food safety requirements. Productivity, yield and selling price are the major factors affecting the economics of SCP production. Microbial inoculants, which are used as a process aid, generally have a higher value. In this case, the objective of the production process is to optimiseyield of viable cells of defined biological activity with good shelf life characteristics.

Saccharomyces cerevisiae, is categorised primarily as a microbial inoculant. Inactive dried brewer's or baker's yeast is also used as a dietary source of vitamins and trace minerals in specific medical conditions. Considerable amounts of yeast extract are produced from bakers yeast as a source of flavour and vitamins.

SINGLE CELL PROTEIN PRODUCTION

Product Safety and Quality

SCP has applications in animal feed, human food and as functional protein concentrates. Some bacterial SCP have amino acid profiles similar to animal/plant protein. Yeast, fungal and soya bean proteins tend to be deficient in methionine. Ingestion of RNA from non-conventional sources should be limited to 50g per day. Ingestion of purine compounds arising from RNA breakdown, leads to increased plasma levels of uric acid, which can cause gout and kidney stones. High content of nucleic acids causes no problems to animals since uric acid is converted to allantoin which is readily excreted in urine. Nucleic acid removal is not necessary from animal feeds but is from human foods. A temperature hold at 64°C inactivates fungal proteases and allows RNA-ases to hydrolyse RNA with release of nucleotides from cell to culture broth. A 30 min stand at 64°C reduces intracellular RNA levels in *Fusarium graminearum* from 80mg/g to 2mg/g.

Production

Large scale fermenters are required High biomass productivity requires high oxygen transfer rates which promotes high respiration rates which in turn increase metabolic heat production and the need for an efficient cooling system. In order to maximise fermentation productivity it is essential to operate continuous fermentation processes. Different processes have adopted different fermenter designs with respect to process requirements.

BP Process – Candida on n-paraffin

Mechanically agitated fully baffled fermenters with turbine mixers. Reactor feed consists of paraffin, gaseous ammonia and other salts. Oxygen requirements per unit biomass produced by aerobic micro-organisms grown on n-hexadecane is 2.5 times higher rhan that required for growth on glucose and amounts to 2.2 kg O_2 per kg biomass. Heat produced was 25000 kJ/kg biomass. Fermenters require substantial agitation. Because of the insoluble nature of alkanes, they exist in agiteated fermenter as suspensions of alkane drops 1-100mm in diameter. Cell recovery is by centrifugation, producing 15% dry solids, evaporation to 25% dry solids and spray drying.

ICI Process – Methylophilus Methylotrophus from Methanol

Pressure recycle fermentor. A combination of an airlift and loop reactor consisting of an airlift column, a down-flow tube with heat removal and a gas-release space. The nitrogen source is ammonia gas and pH is controlled between 6 and 7. Cell specific growth rate is approximately 0.5h-1 and cell yield of 0.5 g/g. Methanol is oxidised via dehydrogenation to formaldehyde which can either be assimilated for conversion to cell mass or further oxidised to CO_2 with concomitant energy production. Cells are recovered by agglomeration followed by centrifugation, flash dried and ground.

Bel Fromageries Process: Kluyveromyces Marxianus from whey

Whey which contains about 5% lactose, 0.8% protein and 0.2-0.6% lactic acid, is used as a substrate. Biomass production requires an aerobic fermentation whereas aeration is minimal for ethanol production. For feed grade biomass, the entire fermentation contents, containing yeast, residual whey proteins, minerals and lactic acid may be recovered. For preparation of food grade material, cells are harvested by centrifugation, washed and dried. Cell yield is 0.45-0.55 g/g based on lactose consumed.

RHM Mycoprotein Process: Fusarium Graminearum from Glucose

Medium components include food grade glucose syrup,

gaseous ammonia, salts, and biotin. Fermentation pH is controlled at 6 by gaseous ammonia addition, fed into the air inlet stream. Cell concentrations are 15-20g/L and a specific growth rate of up to 0.2h-1 is achieved. Following cyclone separation and an RNA reduction step, cells are recovered by rotary vacuum filtration and formulated into a range of products.

POTENTIAL SUBSTRATES FOR SCP PRODUCTION SULPHITE WASTE LIQUOR

Candida utilis has been produced as a protein supplement by fermentation of sulphite waste liquor in Germany during both world wars. More recently a Finnish company developed a fungal SCP production process, the Pekilo Process, to grow *Paecilomyces varioti* using sulphite waste liquor.

CELLULOSE

Cellulose from natural sources and waste wood is an attractive starting material for SCP production because of its abundance. The association of cellulose with lignin in wood makes it somewhat intractable to microbial degradation. Thermal or chemical pretreatment, used in combination with enzymatic hydrolysis, is usually required. Systems using cellulolytic organisms appear to have promise, but economic viability has yet to be achieved.

WHEY

Whole milk whey or deproteinised whey is a carbohydrate source, which creates disposal problems. (High BOD) Problems associated with whey for SCP production are usually insifficient substrate, seasonal supply variations and its high water content (>90%) which makes transport prohibitively expensive. While most organisms do not grow on lactose as a carbon source, strains of the yeast Kluyveromyces marxianus readily grow on lactose.

STARCH

The symba process was developed in Sweden to produce

SCP from potato starch using two yeast strains. *Saccharomyces fibuligera* produces the enzyme necessary for starch degradation enabling co-growth of *Candida utilis*. This process is known as simultaneous saccharification and fermentation.

GLUCOSE

Food grade glucose was the substrate chosen by RHM for production of fungal SCP using *Fusarium graminearum*. The strategy adopted was to take advantage of mycelial fibre content to produce a range of high added value products including meat analogues for human consumption.

HIGHER ALKANES

The original alkane SCP fermentation process, developed by BP in France used 10-20% wax contained in gas oil. Substrate costs were very low, however due to their crude nature, exhaustive processing was required to recover the yeast free of a gas-oil flavour taint. Other alkane based SCP processes were developed in Italy, Japan and Romania but many of them suffered from the problems of potential carcinogenic residues and most of the plants have never run on full capacity or have been closed.

METHANE/METHANOL

Methane was initially considered as a SCP raw material because, as a gas product, purification problems after fermentation would be minimal.

Disadvantages associated with methane-based processes are related to:

- The greater oxygen requirements necessary to fully oxidise methane compared with paraffins,
- The low solubility of methane in water
- The requirement that the fermentation plant be flame proof as methane-oxygen mixtures are highly explosive.

Methane is however easily converted to methanol which requires less oxygen, less fermenter cooling, is highly water soluble and has minimal explosion risks. ICI, which manufactures bulk methanol, chose this substrate for backterial SCP production for animal feed using the trade name Pruteen. The company designed a non-mechanical "pressure cycle fermenter" which uses air for both agitation and aeration in the worlds largest single aerobic fermenter of 3000m^3 capacity.

The process which produces 50-60,000 tonnes SCP per year, using the organism *Methylophilus methylotrophus, was comissioned in 1979-1980,* but has suffered from dramatic increases in methanol prices. The economic difficulties encountered by ICI with animal feed processes lead to a joint venutre with RHM to produce *Fusarium* SCP in the ICI plant.

CHOICE OF MICROORGANISM

The key criteria used in selecting suitable strains for SCP production should consider the following: The substrates to be used as carbon energy and nitrogen source and the need for nutrient supplementation. High specific growth rates, productivity and yields on a given substrate. pH and temperature tolerance. Aeration requirements and foaming characteristics. Growth morphology in the reactor. Safety and acceptability – non pathogenic, absence of toxins. Ease of recovery. Protein, RNA and nutritional composition of the product. Structural properties of the final product. In general, fungi have the capacity to degrade a wider range of complex plant materials, particularly plant polysaccharides. They can tolerate low pH which contributes to reducing fermenter infections. Growth of fungi as short, highly branched filaments rather than in pellets is essential in order to optimise growth rate. Bacteria, in general, have faster growth rates than fungi and grow at higher temperatures, thereby reducing fermenter cooling requirements. Bacterial and yeast fermentations are easier to aerate. In contrast to fungi, which are easily recovered by filtration, bacteria and yeast require the use of sedimentation techniques and centrifugation.

Bacteria, in general produce a more favourable protein composition than yeast or fungi. Protein content in bacterial can range from 60-65% whereas fungi selected for biomass production and yeast have protein contents in the range of 33-

45%. However, associated with the higher bacterial protein levels is a much higher level of nutritionally undesirable RNA content of 15-25% Microorganisms involved in SCP production must be safe and acceptable for use in food. Organisms should be stable genetically so that the strain with optimal biochemical and physiological characteristics may be maintained in the process through many hundreds of generations.

ECONOMICS OF SCP PRODUCTION

The initial reasoning by companies such as BP and ICI to enter SCP production was to produce, at low cost, high value SCP from petroleum, for addition to animal feed. The intention was to replace imported protein additives such as soyabean meal. Factors which contributed to the failure of hydrocarbon SCP to make a major commercial impact uncluded the 1973 dramatic oil price increases, which raised feedstock and energy costs and the lower price increases achieved by agricultural products such as soyabean. When one considers that crude oil prices increased by a factor of 6 in 1973, and that the cost of substrate for SCP processes represents 40-60% of total manufacturing costs, the negative impact on hydrocarbon based SCP processes may be understood.

Agricultural crops, the major competitor to SCP for animal feed, manifest a remarkable ability to respond to market forces and maintain price stability. By producing human protein supplements, the end value of the product can be appreciated, however the production process has to utilise more conventional and costly substrates for production (glucose). These products take advantage of the high fibre content in certain fungi to produce a product that is a good meat analogue. This product is also low in sodium and fat. The product Quorn is a fungal protein.

PRODUCTION OF SINGLE CELL PROTEIN IN SOLID STATE FERMENTATION

Solid state fermentation (SSF) is growth of microorganisms on predominantly insoluble substrate where there is no free liquid. Generally, under combined conditions of low water activity and presence of intractable solid substrate, fungi show luxuriant growth. Mycelial tips of fungi have immense turgor pressure, which assist in their penetration of hard substrate. Hence, proper growth of fungi in SSF gives much higher concentration of the biomass and higher yield when compared to submerged fermentation. The advantage in SSF process is the unique possibility of efficient utilization of waste as the substrate to produce commercially viable products. The process does not need elaborate prearrangements for media preparation. The process of SSF initially concentrated on enzyme production. But presently, there is worldwide interest for single cell protein (SCP) production due to the dwindling conventional food resources.

Materials and Methods

Microorganism

A local isolate of *Aspergillus niger,* from decaying wood in the soil, was used for the experimental studies. The culture was maintained on PDA slants at 28°C. The subculturing was done once in a fortnight.

Substrate Preparation

Oil free rice bran waste was procured from a local rice mill. It was pretreated to remove lignin and to expose the inner cellulose fibers to cellulolytic attack by the organism. The pretreatment process was carried out by adding one liter of 1% sodium hydroxide to 100 g of rice bran waste and autoclaving at 121°C and at 15 psi for 30 min. The pretreated rice bran was allowed to cool and subsequently filtered and washed to neutral pH. Then it was then dried at 60°C in an oven for 12 h. The dried bran was milled and sieved to obtain 50-mesh size and kept ready for SSF process.

Inoculum

Spores were harvested from a week old *A. niger* in five ml of sterile distilled water. Two ml of spore suspension was added to the pretreated rice bran based medium.

Chapter 11

Genetically Modified Food Crops

EVALUATION OF GENETICALLY MODIFIED FOOD CROPS

Genetic modification, otherwise referred to as recombinant DNA (rDNA) technology or gene-splicing, has proven to be a more precise, predictable and better understood method for the manipulation of genetic material than previously attained through conventional plant breeding. To date, agricultural applications of the technology have involved the insertion of genes for desirable agronomic traits (e.g. herbicide tolerance, insect resistance) into a variety of crop plants, and from a variety of biological sources.

Examples include soybeans modified with gene sequences from a *Streptomyces* species encoding enzymes that confer herbicide tolerance, and corn plants modified to express the insecticidal protein of an indigenous soil microorganism, *Bacillus thuringiensis* (Bt).

A growing body of evidence suggests that the technology may be used to make enhancements to not only the agronomic properties, but the food, nutritional, industrial and medicinal attributes of genetically modified (GM) crops.

Regulatory supervision of rDNA technology and its products has been in place for a longer period of time in the United States than in most other parts of the world. The methods and approaches established to evaluate the safety of products developed using rDNA technology continue to evolve in response to the increasing availability of new scientific information.

As our understanding of the potential applications of the technology is broadened, the safety of products developed using rDNA technology and the potential effects of introduced gene sequences on human health or the environment will be more closely scrutinized. In fact, much of the knowledge acquired during the commercialization of the products of rDNA technology in agriculture is now finding application in evaluating the safety of products developed through more conventional means.

The objective of this chapter is to provide the reader with an overview of the significant events leading up to the present, science-based, regulatory framework that exists for the safety evaluation of GM food crops within the United States. An attempt has been made to discuss concerns over the sufficiency of existing regulations, as well as to highlight recent initiatives taken by federal regulatory agencies to address them. Through better communication of how the regulatory process functions within the United States, it is anticipated that current and future applications of rDNA technology in agriculture will be met with a greater level of understanding and acceptance.

HISTORICAL PERSPECTIVE

DNA technology was first developed in the 1970s. The initial response of the scientific community, including members of the National Academy of Science (NAS), to the prospects of rDNA technology, was to postpone any further research involving the technology until the potential risks to human health and the environment could be evaluated.

Researchers attending the International Conference on Recombinant DNA Molecules in 1975, otherwise known as the Asilomar Conference, tried to establish a scientific consensus on how best to self-regulate emerging applications of the technology. The conditions and restrictions that were proposed at this conference have formed the basis by which federal guidelines and policies for rDNA technology research were drafted within the United States.

National Institutes of Health (NIH)

The National Institutes of Health (NIH) was the first federal regulatory agency to publish their interests in evaluating the safety of rDNA technology in 1976, in the form of guidelines for the conduct of research. Because of the uncertainties that existed at the time, all research into the potential applications of rDNA technology was limited to the confines of federally funded labouratories under NIH control. After continued research, and a more careful assessment and monitoring of the risks, a set of less restrictive guidelines was published in 1978.

However, the environmental release of organisms developed using rDNA technology outside the confines of controlled labouratory conditions was prohibited unless otherwise approved by the NIH director. In the early 1980s, the NIH established an rDNA Advisory Committee (RAC) to review all data and experience gained with applications of the technology under its control. Based on recommendations of the RAC, a more relaxed set of research guidelines was published by the NIH in 1983.

The NIH approved the first environmental release of an organism developed using rDNA technology (ice-minus strain of *Pseudomonas*) in 1983. In response, they were criticized for failing to prepare a statement or assessment of the environmental impact of their regulatory decision as required under the National Environmental Policy Act (NEPA). Once the legal controversy had subsided, all responsibility that the NIH had for regulating the environmental introduction of GM organisms was relinquished. Nevertheless, NIH guidelines continue to be referenced in assessing the safety of rDNA research performed within industry, federal and other state labouratories. However, it was unclear which federal regulatory agencies would be responsible for ensuring the safety of the products developed using rDNA technology.

Office of Science and Technology Policy (OSTP)

In response to a need for clarification, the Office of Science and Technology Policy (OSTP) began work on the development of a policy to establish a federal regulatory framework for evaluating the safety of products developed using rDNA technology. Following an opportunity for public comment, OSTP published a final version of the 'Coordinated Framework for Regulation of Biotechnology' (Coordinated Framework) in 1986.

Table. Overview of Responsible Agencies under the Coordinated Framework

Responsible agency	Products regulated	Reviews for safety
FDA	Food, feed, food additives, veterinary drugs	Safe to eat
USDA	Plant pests, plants, veterinary biologic	Safe to grow
EPA	Microbial/plant pesticides, new uses of existing pesticides, novel microorganisms Safe for the environment	Safety of a new use of a companion herbicide

The policy provided the basis by which federal regulatory agencies got involved in evaluating the safety of products at later stages of commercial development at that time.

According to the Coordinated Framework, the products of rDNA technology should be regulated on the basis of the unique characteristics and features that they exhibit, not their method of production. The products of rDNA technology were considered to pose risks to human health and the environment similar to those posed by conventional products already regulated within the United States. As a result, no new federal regulatory agencies or regulations were required. The Coordinated Framework did not, however, rule out the possibility of the development of new guidelines, procedures, criteria or even regulations to supplement or alter the scope of existing statutes for the products of rDNA technology.

The Coordinated Framework identified three federal regulatory agencies within the United States: the US Food and

Drug Administration (US FDA), the US Department of Agriculture (USDA) and the US Environmental Protection Agency (US EPA), as having primary responsibilities for evaluating the products of rDNA technology under development at that time.

In 1992, the OSTP released another document entitled, 'Exercise of Federal Oversight within the Scope of Statutory Authority: Planned Introductions of Biotechnology Products Into the Environment', outlining the proper basis by which federal regulatory agencies were expected to exercise their regulatory authority. As with conventional products, dependent upon the intended use and function, more than one federal regulatory agency may share an interest in evaluating the safety of a product developed using rDNA technology. If more than one federal regulatory agency has an interest, lead agencies are identified as being responsible for coordinating activities to limit any potential duplication of efforts. Although federal regulatory agencies worked independent of one another, it was realised that close working relationships would need to be established in order to evaluate effectively the safety of products developed using rDNA technology.

Recently, the OSTP teamed up with the White House Council on Environmental Quality (CEQ) to perform a six-month inter-agency evaluation of the federal regulatory agency responsibilities in evaluating the environmental safety of products developed using rDNA technology. A case-study approach for a variety of different classes of products developed using rDNA technology was used to evaluate the level of federal regulatory agency involvement, to identify strengths, weaknesses and areas of potential improvement. The review concluded that none of the previously approved products of rDNA technology has had any significant negative impact on the environment.

Although all the case studies were published, OSTP/CEQ failed to reach a consensus on issues relating to the relevant strengths and weaknesses of the existing regulatory structure within the time allotted for the completion of its review. A review of the case studies published provides a comprehensive interpretation of the responsibilities of each federal regulatory agency in ensuring the safety of the products developed using rDNA technology.

National Academy of Sciences (NAS)

The National Academy of Sciences (NAS), and its operating arm, the National Research Council (NRC), have served as a primary source of scientific, technological, human health and environmental policy advice during the development of regulatory approaches for the safety evaluation of products developed using rDNA technology within the United States.

In 1987, the NRC published a report concerning the potential human health and environmental hazards associated with the commercial introduction of GM organisms, entitled *Introduction of Recombinant DNA-Engineered Organisms into the Environment: Key Issues*. The risks associated with the introduction of GM organisms were considered to be essentially the same in kind as those associated with unmodified organisms.

In other words, rDNA technology did not appear to introduce any unique risks as compared to the products that had been developed using more conventional methods of genetic modification. In reaching these conclusions, the NRC performed an evaluation of the similarities and differences in the properties exhibited by products developed using a variety of different techniques. To this day, the conclusions of this report continue to be referenced by the regulators and developers of GM crop varieties worldwide.

A subsequent NRC report, entitled *Field Testing Genetically Modified Organisms: Framework for Decisions*, also reached similar conclusions, but provided additional guidance as to how regulatory decisions concerning the introduction of GM organisms should be made. The NRC recommended that regulatory decisions concerning the introduction of GM organisms should be made on a case-by-case basis. Consistent with the Coordinated Framework, the NRC did not consider the nature of the process used for the genetic modification of an organism to be a useful criterion for determining whether a product requires less or more regulatory oversight. As a result, no valid reason existed to regulate organisms genetically modified via modern techniques (e.g. rDNA technology) any differently from organisms genetically modified via more conventional means. Similar conclusions have been published in the reports of international standards-setting organizations.

In retrospect, within both reports, the NRC acknowledged that modern and conventional methods of genetic modification are not without risks to human health or the environment, and as a result, neither could be considered inherently more risky. As a result, regulatory decisions concerning the safety of the products of rDNA technology need only to take into consideration the specific characteristics exhibited by a particular GM organism and the environment in which it is to be introduced, and not the method by which it has been produced. The NRC has further articulated their conclusions into what is now commonly referred to as the 'Concept of Familiarity'.

Although familiarity with the characteristics of a particular organism or the environment to which it will be introduced would not necessarily mean it was safe, it can be expected to provide a sufficient amount of information to allow for a judgement to be made of the risks. For example, familiarity with a new GM plant variety could be established based on comparisons between characteristics of the parent line or other crop species exhibiting similar traits, as well as through the results of actual field tests involving the GM plant. These principles were further elabourated upon by the Organisation for Economic Co-operation and Development (OECD), and as a result, have been referenced in the development of regulatory policies for evaluating the safety of GM crops on a global basis.

More recently, a committee established by the NRC published a report entitled *Genetically Modified Pest-Protected Plants: Science and Regulation,* based on a review of all scientific and regulatory data collected during the regulatory approval process for GM crops within the United States. The primary objective was to assess independently the effectiveness of existing and proposed regulations for the safety evaluation of GM crops expressing plant pesticides (i.e. plant-incorporated protectants). No new evidence was identified to suggest plants expressing plant pesticides posed any greater risk to human health or the environment as a result of their genetic modification. In fact, the NRC concluded, 'with careful planning and appropriate regulatory oversight, commercial cultivation of GM plants is not expected to pose higher risks and may pose less risk than other commonly used chemical and biological pest-management techniques'.

However, the NRC report included requests for federal regulatory agencies to further strengthen the current regulatory approval process through better coordination and communication between agencies, on-going investment in the research and monitoring of potential human health (e.g. allergenicity) and environmental impacts (e.g. insect resistance), and by providing greater access to information evaluated in support of regulatory decisions.

US FEDERAL REGULATIONS FOR AGRICULTURAL BIOTECHNOLOGY

Regulatory systems for the products of agricultural biotechnology have been in existence since the mid- to late 1980s within the United States. The regulatory approach to the safety evaluation of plants developed using rDNA technology has evolved in the best interests of research scientists, industry and the general public. The agricultural products of rDNA technology, such as GM foods and crops, may require approvals from up to three regulatory agencies; the US FDA, the USDA and the US EPA, depending upon the characteristics exhibited by the GM plant, its proposed use and introduced traits. The same standards of safety are applied to all products regardless of the technology used in their development.

The US FDA is responsible for ensuring the human safety of all new foods and food components, including products developed using rDNA technology, under the Federal Food,

Drug, and Cosmetic Act (FFDCA). The USDA evaluates the potential of a GM plant to become a plant pest following its environmental introduction under the Federal Plant Pest Act (FPPA). The US EPA evaluates pesticides, including plant systems modified to express pesticides (e.g. insect-protected or virus resistance), under the Federal Insecticide, Fungicide, and Rodenticide Act (FIFRA). As a result, the expression of an insecticidal protein in a food crop would undergo review by the USDA, US EPA and US FDA; a GM food crop exhibiting a modified oil content would be evaluated by the USDA and US FDA; and a non-food horticultural plant developed using rDNA technology for any other purpose (e.g. flower colour) would be subject to review by the USDA alone.

In most instances, obtaining all necessary approvals for the commercialization of an agricultural crop developed using rDNA technology takes a decade or more. However, the exact amount of time required will depend on the need to confirm performance, to evaluate characteristics of the food, environmental effects, and to produce the required amount of seed before the product can be distributed and commercially grown by farmers. Up to five years of field trials (5-10 generations of plants) are required for the developer of a new plant variety to collect sufficient data to meet the reporting requirements of the USDA. An additional five months to two years may be required for the US FDA, USDA and/or US EPA to complete all necessary product consultations, reviews and approvals.

Approval for the first commercial planting of a GM food crop was not issued until 1995. Since 1995, more than 40 new agricultural crops developed using rDNA technology have received approval for commercial planting within the United States. In 1999, approximately 72% of the total 39.9 million hectares (more than 98 million acres) of GM crops grown worldwide were planted in the United States. Herbicide-tolerant soybeans (54%), Bt corn (19%) and herbicide tolerant canola (9%) accounted for approximately 82% of the GM plants cultivated. With an increasing number of agricultural biotechnology products reaching later stages of commercial development, it is anticipated that the overall area planted with GM crops will continue to rise.

The regulatory approach to evaluating the human health and environmental safety of GM crops within the United States is best described as a science-based, case-by-case assessment of hazards and risks. This approach has provided the flexibility required to reduce the regulatory burden placed on products that have been determined to be of low risk or concern. All agencies involved in the regulation of plants developed using rDNA continue to implement and develop policies based on recommendations made within the Coordinated Framework. In the future, the US FDA, USDA and the US EPA will be dedicating additional resources towards communicating how GM food and food components are regulated within the United States, and how these regulations function to be protective of both human health and the environment.

FOOD AND DRUG ADMINISTRATION (US FDA)

Authority

The US FDA is responsible for ensuring the safety and wholesomeness of all food and food components, including the products of rDNA technology, under the FFDCA. The US FDA has the authority for the immediate removal of any product from the market that poses potential risk to public health or that is being sold without all necessary regulatory approvals. As a result, a legal burden is placed on developers and food manufacturers to ensure the commodities utilized and foods available to consumers are safe and in compliance with all legal requirements of the FFDCA. In order to understand the regulatory approach followed by the US FDA in the safety evaluation of GM crops, it is useful to consider food and food safety from a historical context.

People had been consuming foods derived from agricultural crops for many years prior to the existence of any food laws or regulations within the United States. Based on this experience, agricultural crops have been accepted as being safe for consumption as food, without additional testing to demonstrate their safety. As long as the new crop variety has exhibited similar agronomic properties, and an appropriate taste and appearance, it has been considered safe to consume. As a result, most foods consumed today, in particular whole foods (i.e. fruits, grains and vegetables) and conventional foods, have not been subject to any kind of premarket review or approval by the US FDA.

Nevertheless, food scientists have a good understanding that many of the commonly consumed agricultural crops contain natural toxicants (e.g. tomatine in tomatoes, solanine in potatoes, cucurbiticin in cucumber, psoralens in celery, etc.). As a result, new plant varieties may be subject to routine chemical analyses to ensure that none of these substances is present at potentially harmful levels. This type of general approach has been used in assessing the safety of thousands of new plant varieties that have been developed over a number of decades of crop breeding without compromising the safety of whole foods.

Role

Consistent with recommendations within the Coordinated Framework, the US FDA considered existing provisions of the FFDCA to be sufficient for the regulation of foods and food components developed using rDNA technology. It was concluded that the scientific and regulatory issues posed by the products of rDNA technology were not significantly different from those posed by conventional products. As a result, GM foods and food components have been subject to the same standards of safety as already exist for the regulation of other foods and food components under the FFDCA. In order to better communicate interpretations of existing provisions of the FFDCA as they relate to the safety evaluation of foods derived from new plant varieties, including the products of rDNA technology, the US FDA released a policy statement in 1992 entitled 'Statement of policy: foods derived from new plant varieties'.

The US FDA has considered the use of genetic modification (i.e. rDNA technology) in the development of new plant varieties to represent a continuum of conventional plant breeding practices (e.g. mutagenesis, hybridization, protoplast fusion, etc.), and as a result, the safety evaluation of all new plant varieties, not just those developed using rDNA technology, have been evaluated based on an objective analysis of the characteristics of a food or its components, and not on its method of production.

Definition and Scope of Bioengineered Foods

The recently proposed rule of the US FDA concerns 'bioengineered foods' which have been defined as 'foods derived from plant varieties that are developed using *in vitro* manipulations of DNA (generally referred to as rDNA technology)'. As a result, the proposed rule has a much narrower focus than the 1992 US FDA Statement of Policy.

The US FDA has explained the need for a change in emphasis based on their expectations that many of the new plant varieties exhibit a greater potential to 'contain substances that are significantly different from, or that are present in food at a significantly different level than before'. As a result, the substances present in foods and food components derived from new plants developed using rDNA technology are less likely to be considered GRAS, and as a result, will require pre-market approval from the US FDA.

Safety and Nutritional Evaluation

The primary objective of the safety and nutritional evaluation is to demonstrate that the food derived from a new plant variety is as safe or nutritious as foods already consumed as a part of the diet. For new plant varieties, including those developed using rDNA technology, a science-based approach is used to focus the evaluation on the demonstrated characteristics of the food or food component. The evaluation of a GM food or food component typically involves reviewing information or data on any newly introduced substances, the known levels of toxicants, as well as the nutritional composition of the plant following modification. Substances that raise safety concerns (e.g. toxicants, allergens) would be subject to more extensive evaluation, since both intended and unintended changes may affect the levels of toxicants and nutrients in a food following the modification.

Guidance for performing a safety and nutritional evaluation was provided in the 1992 US FDA Statement of Policy, in a series of flow charts and text that cover:

- The crop that has been modified;
- Source(s) of the introduced genetic material;
- New substances intentionally added to the food as a result of the genetic modification (e.g. proteins, but also fatty acids, and carbohydrates).

Documentation required to support the evaluation typically includes: the purpose or intended technical effect of the modification on the plant, together with a description of the various applications or uses, a molecular characterization of the modification including the identities, sources and functions of introduced genetic material; information on the expressed protein products encoded by introduced genes; information relating to the known or suspected allergenicity and toxicity of any expressed gene products; for foods known to cause allergy, information on whether the endogenous allergens have been altered by the genetic modification; information on the compositional and nutritional characteristics of the foods, including anti-nutrients; and in some instances, comparative results of feeding studies involving the foods derived from plants modified using rDNA technology and the non-modified counterpart.

In performing its evaluation, the US FDA is particularly interested in the identification of inherent toxicants, known or potential allergens, assessing the concentration and bioavailability of essential nutrients, the safety and nutritional value of any newly introduced proteins, and the identity, composition, and nutritional value of modified carbohydrates, fats and oils.

If additional questions of safety remain following this evaluation, further toxicological studies may need to be performed. It is recognized that absolute assurance of the safety of any food does not exist. As a result, the goal of the safety evaluation is to establish a reasonable certainty of no harm under anticipated conditions of consumption. With this in mind, experience with the existing food supply has provided the basis for evaluating the safety of new food or food components. Both the Food Advisory Committee and Committee for Veterinary Medicine have been extensively involved in the development of approaches for the safety and nutritional evaluation of foods and food components derived from new plant varieties, including those developed using rDNA technology.

PRODUCT CHARACTERIZATION

Product characterization takes into consideration information relating to the modified food crop, the introduced genetic material and its expression product, and acceptable levels of inherent plant toxicants and nutrients. All characteristics of the gene insert must be known, including the source(s), size, number of insertion sites, promoter regions, and marker sequences. It must be established that the transferred genetic material does not come from a pathogenic source, a known source of allergens, or a known toxicant-producing source. The introduced genetic material should be well characterized to ensure that the introduced gene sequences do not encode harmful substances and are stably inserted within the plant genome to minimize any potential opportunity for undesired genetic rearrangement.

Analytical data are required to evaluate the nutritional composition, the levels of any known toxicants, anti-nutritional and allergenic substances, and the safety-in-use of antibiotic resistance marker genes. Any new substances introduced into crops through rDNA technology (e.g. proteins, fatty acids, carbohydrates) will be subject to pre-market review as food additives by the US FDA, unless substantially similar to substances already safely consumed as a part of foods, or that are considered GRAS. To date, substances that have been added to foods through rDNA technology have been previously consumed or have been determined to be substantially similar to substances already consumed as a part of the diet. As such, introduced substances have been considered exempt from the requirement for pre-market approval as food additives with the US FDA.

A more rigorous safety evaluation of a GM crop is warranted if the introduced gene sequence(s) has not been fully characterized, the nutritional composition has been significantly altered, antibiotic resistance marker genes have been used during its development, or if an allergenic protein or toxicant has been detected at levels higher than what is typically observed in edible varieties of the same crop species. In any event, determinations as to the safety of substances that have been introduced into new plant varieties through rDNA technology are made on a case-by-case basis.

Although an evaluation of the introduced gene sequences and expression product(s) provides assurance as to their safety, further studies may be required to predict whether unexpected effects may result following their interaction with other genes within the plant. In addition, the product characterization of a GM plant involves assessing sequence homology to known toxicants and allergens, thermal and digestive stability, and if required, the results of both *in vitro* and *in vivo* assays to demonstrate lack of toxicity.

COMPOSITIONAL ANALYSIS

The results of field trials performed over several years serve to characterize the phenotypic and agronomic characteristics exhibited by the plant (e.g. height, colour, leaf orientation, susceptibility to disease, root strength, vigour, fruit or grain size, yield, etc.), as well as to provide the materials required for the compositional analysis. Any anomalies in the phenotypic or agronomic characteristics exhibited by a plant may result in a requirement for additional information. Protein, fat, fibre, starch, amino acid, fatty acid, ash and sugar levels are determined, as well as the levels of anti-nutrients, natural toxicants or known allergens. Studies of the nutritional composition are performed to determine whether the levels of any key nutrients, vitamins or minerals have been altered as a result of the genetic modification.

Based on the results of these studies, a determination is made as to whether the phenotypic and agronomic characteristics of a GM crop or the concentrations of inherent constituents fall within ranges typical of its conventional counterpart. If inserting a new gene causes no change in any of the assessed parameters, the US FDA can conclude with reasonable assurance that the GM crop is as safe as the conventional crop. If the levels of essential nutrients or inherent toxicants are found to be significantly different in the GM crop, the US FDA may recommend additional action prior to commercialization, such as obtaining food additive status, or the use of specific labels to alert consumers of an altered nutritional content, etc.

ALLERGENICITY

In consultation with scientific experts in the areas of food safety, food allergy, immunology, biotechnology and diagnostics, the US FDA published guidelines for assessing the allergenicity of GM foods or food components in 1994.

The approach to assessment is multi-faceted, incorporating data regarding the origin of the genetic material, and the biochemical, immunological and physicochemical properties of the expressed protein. The overall assessment is reliant upon the fact that all known food allergens are proteins and, notwithstanding the number of shared properties between allergenic and non-allergenic food proteins, food allergens tend to exhibit a number of similar characteristics. In general, food allergens share a number of common properties: they have a molecular weight of over 10 000 Da; they represent more than 1% of the total protein content of the food; they demonstrate resistance to heat, acid treatment, proteolysis and digestion; and they are recognized by IgE.

For gene sequences derived from known allergenic sources (e.g. peanuts), the developers of GM plants are expected to demonstrate that allergenic proteins have not been introduced into the food. For assessment purposes, it is assumed that any genetic material derived from a known allergenic source will encode for an allergen. To demonstrate otherwise, the amino acid sequence of an expressed protein must be compared with that of known allergens using protein sequence databases. Furthermore, *in vitro* and/or *in vivo* immunologic analyses using the sera of allergic patients sensitive to the source of the genetic material may need to be performed to determine whether or not a potentially allergenic protein is being expressed in the GM food.

Some GM foods may be modified to express genes from a source that is not known to be allergenic when consumed. Under these circumstances, the US FDA follows a similar decision tree-based approach to determining the allergenic potential of the expression product. In assessing these proteins, should any amino acid sequence exhibit homology with a known allergen, the expressed protein would then be evaluated in immunologic tests using the sera of patients known to be allergic to the identified homologous protein. Regardless of the origin of the genetic material, physicochemical studies are performed *in vitro* to provide information concerning the expected stability of the expressed protein. All known food allergens tend to be resistant to digestive degradation, as demonstrated in simulated gastric fluid models, or to decomposition under conditions of food processing.

In making a determination regarding a GM food, it is the totality of the biochemical, immunological and physico-chemical properties of the introduced protein that provides guidance as to the allergenic potential of such a protein being expressed in food. The level of protein expressed, produced and consumed as a part of the diet is also a primary indicator of the allergenic potential, since nearly all food allergens are known to be major proteins in their respective foods. If the results of any of these studies suggest an allergenic potential, the US FDA may recommend further scientific evaluation, require special labelling to alert sensitive consumers, or alternatively caution the developer about proceeding with the development of a particular GM food.

Most recently, a joint FAO/WHO Expert Consultation on Allergenicity of Foods Derived from Biotechnology recommended a revised decision tree. This decision-tree strategy was modified from the previous version FAO/WHO to include a revised definition of sequence homology for gene product comparison; a greater emphasis on serum testing, even with gene products without homology to known allergens and not derived from an allergenic source; and animal models to assess potential allergenicity, despite acknowledgement by the Expert Consultation that these models are currently under development and at present, not predictive of food allergies in humans.

More recently, the *ad-hoc* Open-Ended Working Group on Allergenicity established by the *ad-hoc* Intergovernmental Codex Task Force on Foods Derived from Biotechnology, considered the FAO/WHO strategy in drafting an approach to assessing the potential allergenicity of foods derived from rDNA plants.

The Codex Working Group recognized the absence of a definitive predicitve test for allergenicity in humans to a newly expressed protein and recommended an integrated, stepwise assessment strategy. The strategy recommended by the Working Group, and accepted by the Codex Task Force for inclusion in the draft forwarded for final adoption by the Codex Alimenarius Commission, is consistent with that of FAO/WHO. However, the Working Group suggested that some of the modifications included in FAO/WHO could contribute to the overall weight of evidence of any conclusion of potential allergenicity (e.g., allergen specific serum depositories, animal models), pending development and validation.

ANTIBIOTIC RESISTANCE

The use of antibiotic-resistance genes as selectable markers has been common practice in the development of new plant varieties using rDNA technology. Concerns relate to the potential transfer of antibiotic-resistance genes from GM plants to pathogens in the environment or to the gut of humans consuming foods or food components. Issues relating to the use of antibiotic resistance genes were identified in the 1992

US FDA Statement of Policy Statement as well as discussed in additional guidance entitled *Guidance for Industry: Use of Antibiotic Resistance Marker Genes in Transgenic Plants*. The guidance provided within these documents was established in consultation with experts in the fields of microbiology, medicine, food safety, bacterial and mycotic diseases, and includes suggestions with respect to the continued safe use of antibiotic-resistance marker genes by the developers of new plant varieties.

The use of marker genes that encode resistance to clinically important antibiotics has raised questions as to whether their presence in food could reduce the effectiveness of oral doses of the antibiotic or whether the gene present in the DNA could be transferred to pathogenic microbes, rendering them resistant to treatment with the antibiotic. The risk of transfer of antibiotic-resistance genes from plants to microorganisms considered to be pathogenic to humans, however, is considered to be minimal if not insignificant. Furthermore, the potential risks are becoming less of a concern as more developers are beginning to research the use of alternative technologies (e.g. non-resistance-based markers) in plant breeding. The conclusions with respect to the safe use of antibiotic-resistance marker genes are consistent with the findings of other national and international food safety organizations.

Consultation and Filing Process

The submission of a Pre-market Biotechnology Notification (PBN) has recently been proposed as a mandatory requirement for the commercialization of bioengineered foods and food ingredients within the United States. Minimal differences exist between the information to be submitted as a part of a PBN and that previously presented in voluntary consultations with the US FDA. As proposed, developers are still being encouraged to consult with the US FDA as early and as often as necessary in the development of a bioengineered food, such that any potential scientific or regulatory concerns can be identified and addressed prior to the submission of the PBN.

Guidelines for performing consultations with the US FDA were released in a publication entitled *Guidance On Consultation Procedures: Foods Derived from New Plant Varieties* in 1997. The publication recommended an approach for developers to proceed by submitting a request for consultation, outlined the internal process by which all requests would be handled, and included additional guidance as to the type of safety and nutritional information to be presented during consultation with the US FDA.

Once sufficient safety and nutritional information has accumulated to demonstrate that a product is safe and in compliance with the FFDCA, developers typically schedule a consultation to present their scientific findings and conclusions to the US FDA. Consultations prior to notification not only serve to keep the US FDA informed of advances made in the application of rDNA technology in food production, but also keep the developers of bioengineered foods aware of emerging safety, nutritional or regulatory concerns of the US FDA.

Consultations are considered complete when all safety and regulatory concerns between the US FDA and the developer have been resolved. The US FDA proposes to perform an initial evaluation of a PBN within 15 days of receipt to determine completeness, at which point, if considered complete, the PBN will be filed, and a response can be expected within 120 days. The US FDA does not issue a product approval *per se*, but informs the developer by letter that:

- The evaluation period has been extended;
- The notice is not complete and why;
- It has no further questions 'at this time' based on the information that has been presented.

Labelling

The FFDCA defines what information must be disclosed to consumers on a food label, such as the common or usual name, and other limitations concerning the representations or claims that can be made or suggested about a food product. All foods must be labelled truthfully and not be misleading to consumers. Taking this into consideration, the FFDCA does not stipulate the disclosure of information on the basis of consumer desire to know. Labelling may be considered misleading if it fails to reveal material facts in light of representations that are made with respect to a product.

The labelling of foods derived from new plant varieties, including plants developed using rDNA technology, was originally addressed in the 1992 US FDA Statement of Policy, and most recently discussed in Draft Guidance for Industry for the voluntary labelling of bioengineered foods.

To date, the US FDA is not aware of any information that would distinguish foods developed using rDNA technology (e.g. bioengineered foods) as a class from foods developed through other methods of conventional plant breeding, and as such, have not considered the method of development a material fact requiring disclosure on product labels. Nevertheless, after extensive consultation, including thousands of written comments and a series of public meetings, the US FDA has observed 'a general agreement that providing more information to consumers about bioengineered foods would be useful'.

Requirements for Labelling

Special labelling is required if the composition of the bioengineered food differs significantly from its conventional counterpart. For example, for a food that has been genetically modified to contain a new major sweetener, a new common or usual name or other labelling may be required. Similarly, if a GM food contains an allergen that consumers would not expect to be present in that food, special labelling may be necessary to alert sensitive consumers. If a protein commonly associated with an allergic reaction (e.g. peanut protein) is transferred to another food through genetic modification, the US FDA would evaluate whether labelling would provide sufficient consumer protection. If labelling would not be considered to provide a sufficient level of protection, the US FDA would take appropriate steps to ensure the GM food would not be marketed. Therefore, current policy requires a GM food to be labelled when the resulting product poses a safety issue or is substantially different from its conventional counterpart, and as a result, could be considered to pose a misrepresentation to consumers.

The 1992 US FDA Statement of Policy and the recent Draft Guidelines do not consider the use of rDNA technology in the development of food products to be a material fact requiring specific disclosure on the label. Rather it is a method of development, similar to other methods of plant breeding, which have not required disclosure on the label. Bioengineered foods cannot be distinguished compositionally from foods modified through more conventional methods and thus do not require specific disclosure through labelling. The Draft Guidelines reaffirm the US FDA position that bioengineered foods do not require special labelling.

Voluntary Labelling

To provide guiding principles for voluntary labelling, in recognition of the desire of certain manufacturers to label foods as produced either with or without bioengineering, the US FDA published a 'Draft Guidance for Industry: voluntary labelling indicating whether the foods have or have not been developed using bioengineering'. Emphasizing that the use of rDNA technology 'is not a material fact', the US FDA recognizes that some consumers want disclosure of bioengineered content and that some manufacturers wish to provide it. In response, Draft Guidance was issued with suggestions concerning the use of labelling statements that are not considered misleading.

PRESENCE OR USE OF RDNA

The US FDA provided several examples of how disclosure of bioengineering can be accomplished, be informative and not be misleading.

- *Example* 1: 'Genetically engineered' or 'This product contains corn meal that was produced using biotechnology'. These disclosures reveal the minimum amount of optional information about bioengineering.
- *Example* 2: 'This product contains high oleic acid

soybean oil from soybeans developed using biotechnology to decrease the amount of saturated fat.' This statement explains the proper and required name of this type of soybean oil, since it differs from what would be considered as soybean oil. The optional comments about biotechnology and decreasing saturated fat provide information that could be seen as benefits of the product.

- *Example* 3: 'These tomatoes were genetically engineered to improve texture.' This example was one included to illustrate how it could be misleading to consumers if they cannot discern a difference in texture, but would not be misleading if they can tell a difference. If the former, and the new texture is to facilitate processing, then this intention should be made clear, i.e. 'to improve texture for processing'.
- *Example* 4: 'Some of our growers plant tomato seeds that were developed through biotechnology to increase crop yield.' This is another example of optional information that would explain an indirect, agricultural benefit.

ABSENCE OR NOT BIOENGINEERED

The US FDA provides an important commentary in the Draft Guidance about 'genetically modified organisms (GMO)' versus 'bioengineered'. It would be technically inaccurate to use the phrase 'not genetically modified' or 'GMO-free' to mean that bioengineering was not used.

This is because most conventional foods have been genetically modified over the years by traditional crop breeding practices.

Examples of acceptable voluntary statements would include:

- We do not use ingredients that were produced by biotechnology.'
- This oil is made from soybeans that were not genetically engineered.'
- Our tomato growers do not plant seeds developed using biotechnology.'

Another important point is made about a term such as 'GMO-free' that is misleading for two reasons:

- Many foods do not contain organisms anyway and therefore should not be labelled as 'organism-free';
- Free' implies complete absence or 'zero' amount of bioengineered material.

In a practical sense, it is impossible to demonstrate analytically the complete absence of anything. Therefore, in reality, a threshold for possible adventitious presence of a low level of bioengineered material may be necessary and has been the subject of much debate. The Agency also provides additional guidance about misleading statements that:

- Could be interpreted to suggest that the absence of bioengineering would make the food superior to the bioengineered alternative;
- Claim the absence of one bioengineered ingredient when the food contains another ingredient that is bioengineered;
- Claim that a food is not bioengineered when in fact this type of food (e.g. green beans) has never been modified through rDNA technology.

DEPARTMENT OF AGRICULTURE

The Animal and Plant Health Inspection Service (APHIS) of the USDA is responsible for the protection of domestic agricultural resources and the environment from the threats posed by pests and diseases. The Federal Plant Pest Act (FPPA) has been the primary statute adapted by APHIS in regulating the products of agricultural biotechnology. Under the FPPA, APHIS has the authority to restrict the import of plant pests, to introduce measures (e.g. quarantines) to prevent their movement, or to destroy plant pests cultivated in violation of the FPPA. The requirements under the FPPA are equally applicable to scientists within company, academic, research, and private organizations.

Definition of Plant Pests

A plant pest is any agent, substance or organism, or parts

thereof, that may cause injury, disease or damage, either to plants or to the environment in which it is introduced. APHIS maintains a list of organisms considered to be plant pests, and therefore, subject to regulation under the FPPA. If an organism is not on the list, it may still be subject to regulation as a plant pest if there is reason to believe it is or will act as a plant pest. Organisms considered plant pests, or that exhibit the potential to be plant pests, are treated as 'regulated articles' under the FPPA.

The regulations extend to the introduction of GM organisms meeting the definition of a plant pest or for which there is reason to believe they are plant pests. APHIS considers the environmental release of a GM crop to be equivalent to the introduction of new organisms. Therefore, until proven otherwise, a GM crop is considered a 'regulated article' under the FPPA. The use of the term 'plant pest' in reference to GM plants means the 'non-pest' nature of the crop has yet to be determined. To date, all field tests conducted have demonstrated that GM crops exhibit no more 'pest-like' traits than their conventional counterparts.

All GM plants developed to date have involved the use of at least one designated 'plant pest' as either promoters or vectors, and as a result, have been subject to regulatory review under the FPPA. APHIS takes the position that an entire plant can be designated as a regulated article, even if it was not developed using plant pests, if the plant is the product of genetic engineering, and the agency determines or has reason to believe it is a plant pest. Given this position, APHIS would likely challenge any attempt to introduce a GM plant into commerce unless the developer has first gone through the APHIS regulatory review process.

Requirements for Commercialization

A number of years of field testing are required to evaluate the agronomic and product quality characteristics exhibited by new plant varieties developed within a labouratory or greenhouse. APHIS regulations outline procedures for obtaining a permit or providing notification prior to the importation, interstate movement or release of a regulated article, which of course, includes field testing.

Prior to an introduction, a proponent must provide notification to APHIS of its intentions or submit an application to obtain a permit. The notification and permit requirements have evolved as an extension of the long-standing programm for the regulation of plant pests within the United States. Since 1987, APHIS has issued permits or acknowledged the receipt of notifications for the field release of more than 6500 new crop varieties.

Permits

For APHIS approval, developers must demonstrate that a new plant variety poses no significant risk to other plants in the environment or is at least as safe as similar crop varieties. APHIS review and approval are required for the shipment of seed and for the conduct of field trials involving new crop varieties. APHIS issues site-specific permits for field tests or releases into the environment.

Prior to field testing, APHIS may seek further clarification of study objectives, specify how, when or where research may be conducted, establish the data to be generated, collected and reported, stipulate additional requirements for the monitoring or securing of test sites, and specify methods for the disposal of crop residues that remain at the end of a study. The results of field trials are often used in making a determination as to whether a GM plant exhibits any effects on non-target species, or poses any unique plant pest problems.

For field trials conducted under either notification or permit, developers of GM plant varieties are required to submit reports to APHIS that include: details regarding the methods of observation; the data collected; and interpretations concerning the effects on plants, non-target organisms, or the environment.

In the event that a GM crop is likely to be the subject of a petition for a determination of non-regulated status, developers must provide a description of known and potential differences, and substantiate that the regulated article is not likely to pose a greater plant risk than the organism from which it was derived.

An environmental assessment is a requirement of the NEPA, Council on Environmental Quality regulations, and USDA procedures for issuing permits under the FPPA. Under the permit process, developers of GM plant varieties must disclose information concerning the plant, test facilities, and control measures in place for its transport and field testing. Based on this information, the USDA performs an assessment of the potential environmental impact following a release. If the agency reaches a 'Finding of No Significant Impact' (FONSI), a permit may then be issued.

Notifications

Depending on the plant species and intended introduction, applying for a permit from APHIS can be a complicated and time-consuming process. In the spring of 1993, APHIS introduced a simplified notification process as an alternative to applying for a permit, which in most instances applies to the introduction of GM plants. To qualify for notification, a GM plant must be introduced in accordance with the eligibility criteria.

Originally, six GM crops (corn, cotton, potatoes, soybean, tobacco and tomatoes) qualified for notification; however, the process was more recently expanded to include most other crop species not capable of becoming a noxious weed. The developers of new GM plant varieties are encouraged to contact APHIS when making a determination as to whether a GM plant would qualify for notification.

If a particular GM plant does not qualify for notification, a more involved regular permit process must be followed. For a GM plant to qualify for notification, it and its predecessor must not be a noxious weed, the introduced gene sequence(s) must be stably integrated within the host, not originate from a plant or animal pathogen, exhibit a known function that will contribute to plant disease, or encode for substances that could be potentially toxic or infective to non-target organisms, such as pharmaceuticals, or viral components other than those coat proteins already established as safe. Performance standards also exist for the conduct of field trials to ensure that an adequate level of containment is provided for all introductions. It is the responsibility of the developer to ensure that an adequate level of containment is provided for field trials involving GM plants under notification. The performance standards utilized will vary according to the biology of the plant and the nature of the introduction. In general, this includes specifying methods for the proper handling, shipment, field monitoring and disposal of plant material at the end of the study.

As a part of the notification process, an applicant must provide information about the plant, identify the source of any genes used; the method of genetic modification; and the size, date and location of any proposed field test or introduction. Notification is required at least 10 days prior to the interstate movement or 30 days in advance of the field testing or importation of a regulated article that qualifies for notification with APHIS.

Index

A

Aseptic Technique 19

B

C

D

E

F

G

H

I

K

L

M

N

O

P

R

S

T

U

V

W